AF739172

CRC Series in Mathematical Models in Microbiology

Editor-in-Chief

Michael J. Bazin, Ph.D.

Microbial Population Dynamics

Editor

Michael J. Bazin, Ph.D.

Soil Microbiology: A Model of Decomposition and Nutrient Cycling

Author

O. L. Smith, Ph.D.

Physiological Models in Microbiology

Editors

Michael J. Bazin, Ph.D.

James I. Prosser, Ph.D.

Physiological Models in Microbiology

Volume II

Editors

Michael J. Bazin, Ph.D.
Senior Lecturer
Department of Microbiology
King's College
London, England

James I. Prosser, Ph.D.
Lecturer
Department of Genetics and Microbiology
University of Aberdeen
Aberdeen, Scotland

CRC Series in Mathematical Models in Microbiology

Editor-in-Chief
Michael J. Bazin, Ph.D.

CRC Press
Taylor & Francis Group
Boca Raton London New York

CRC Press is an imprint of the
Taylor & Francis Group, an **informa** business

First published 1988 by CRC Press
Taylor & Francis Group
6000 Broken Sound Parkway NW, Suite 300
Boca Raton, FL 33487-2742

Reissued 2018 by CRC Press

© 1988 by CRC Press, Inc.
CRC Press is an imprint of Taylor & Francis Group, an Informa business

No claim to original U.S. Government works

This book contains information obtained from authentic and highly regarded sources. Reasonable efforts have been made to publish reliable data and information, but the author and publisher cannot assume responsibility for the validity of all materials or the consequences of their use. The authors and publishers have attempted to trace the copyright holders of all material reproduced in this publication and apologize to copyright holders if permission to publish in this form has not been obtained. If any copyright material has not been acknowledged please write and let us know so we may rectify in any future reprint.

Except as permitted under U.S. Copyright Law, no part of this book may be reprinted, reproduced, transmitted, or utilized in any form by any electronic, mechanical, or other means, now known or hereafter invented, including photocopying, microfilming, and recording, or in any information storage or retrieval system, without written permission from the publishers.

For permission to photocopy or use material electronically from this work, please access www.copyright.com (http://www.copyright.com/) or contact the Copyright Clearance Center, Inc. (CCC), 222 Rosewood Drive, Danvers, MA 01923, 978-750-8400. CCC is a not-for-profit organization that provides licenses and registration for a variety of users. For organizations that have been granted a photocopy license by the CCC, a separate system of payment has been arranged.

Trademark Notice: Product or corporate names may be trademarks or registered trademarks, and are used only for identification and explanation without intent to infringe.

Library of Congress Cataloging-in-Publication Data

Physiological models in microbiology

Bibliography: p.
Includes index.
1. Micro-organisms—Physiology—Mathematical models. 2. Biological models. I. Bazin, Michael J. II. Prosser, James Ivor.
QR84.P46 1988 576'. 11'0724 87-18330
ISBN 0-8493-5953-8 (set)
ISBN 0-8493-5954-6 (v. 1)
ISBN 0-8493-5955-4 (v. 2)

A Library of Congress record exists under LC control number: 87018330

Publisher's Note
The publisher has gone to great lengths to ensure the quality of this reprint but points out that some imperfections in the original copies may be apparent.

Disclaimer
The publisher has made every effort to trace copyright holders and welcomes correspondence from those they have been unable to contact.

ISBN 13: 978-1-315-89654-0 (hbk)
ISBN 13: 978-1-351-07564-0 (ebk)

Visit the Taylor & Francis Web site at http://www.taylorandfrancis.com and the
CRC Press Web site at http://www.crcpress.com

MATHEMATICAL MODELS IN MICROBIOLOGY

M. J. Bazin, Editor-in-Chief

This multivolume series will contain a selection of authoritative articles on the application of mathematical models to microbiology. Each volume will be devoted to a specialized area of microbiology and topics will be presented in sufficient detail to be of practical value to working scientists. A sincere attempt will be made to make the material useful for microbiologists with only moderate mathematical training. Under each title a variety of modeling techniques will be included, and both purely scientific and applied aspects of the subject will be covered.

The objectives of the series will be to introduce microbiologists familiar with the modeling approach to new models, methods of model construction, and analytical techniques, and to encourage those with limited mathematical backgrounds to incorporate modeling as an integral part of their research programs.

PREFACE

Volume II

The multi-volume series on **Mathematical Models in Microbiology** presents the application of mathematical models to the study of microbial processes and interactions. *Physiological Models in Microbiology* consists of two volumes. Volume I considers models of basic growth processes and the effects of environmental factors on microbial growth. Volume II describes models of secondary processes, in particular, microbial death, spore germination, chemotaxis, and surface growth.

Chapter 7 provides a treatment of the kinetics of micorbial death resulting from heat, irradiation, and chemical inactivation and assesses the degree to which a general model may be applied to microbial death. Chapter 8 describes models for the progress of spore germination through several stages to the formation of a vegetative cell. A quantitative analysis of chemotaxis is then provided in Chapter 9, with particular reference to oxygen chemotaxis. Finally, the growth of microorganisms on solid surfaces is discussed in Chapters 10 and 11. In the former, the kinetics of cell adsorption and attachment are considered, while Chapter 11 provides models for development of complex biofilms, their metabolism, and processes involved in their removal.

THE EDITORS

Michael J. Bazin, Ph.D., is a Senior Lecturer in the Department of Microbiology, King's College, University of London.

Dr. Bazin trained as a teacher at St. Luke's College, Exeter, and taught in secondary schools in England and the United States. He received his Ph.D. in Zoology from the University of Minnesota in 1968 after which time he pursued a postdoctoral traineeship in biomathematics at the University of Michigan.

Dr. Bazin has had wide research interests ranging from sexuality in blue-green algae to ethnic differences in skinfold thickness. His current major interests revolve around the application of mathematics to problems in biology and are directed chiefly towards theoretical biology and biotechnology.

James I. Prosser, Ph.D., is a Lecturer in the Department of Genetics and Microbiology, University of Aberdeen, Aberdeen, Scotland. Dr. Prosser received his B.Sc. degree in Microbiology from Queen Elizabeth College, University of London in 1972, and his Ph.D. from the University of Liverpool, Department of Botany in 1975. After three further years in the latter Department, as a Natural Environment Research Council Postdoctoral Fellow and Senior Demonstrator, he took up his current post at the University of Aberdeen.

Dr. Prosser's research interests fall into two main areas. The first is an investigation into the environmental factors affecting growth, activity, attachment, and inhibition of soil nitrifying bacteria. The second is the relationship between growth, branching, and secondary metabolite production by filamentous fungi and actinomycetes. The link between these two areas is the application of a quantitative approach and the use of theoretical models in the study of microbiological processes.

CONTRIBUTORS

Volume I

Arthur C. Anthonisen, Ph.D.
Consulting Engineer
MONTECO
Montgomery, New York

H. Kacser, Ph.D.
Department of Genetics
University of Edinburgh
Edinburgh, Scotland

Prasad S. Kodukula, Ph.D.
Union Carbide Corporation
Technical Center
South Charleston, West Virginia

Lee Yuan Kun, Ph.D.
Department of Microbiology
National University of Singapore
Singapore, Republic of Singapore

Y.-K. Lee, Ph.D.
Department of Microbiology
National University of Singapore
Singapore, Republic of Singapore

T. A. McMeekin, Ph.D.
Reader in Microbiology
Department of Agricultural Science
University of Tasmania
Hobart, Tasmania, Australia

Marcel Mulder, Ph.D.
Department of Biochemistry
B.C.P. Jansen Institute
University of Amsterdam
Amsterdam, The Netherlands

June Olley, D.Sc., Ph.D.
Senior Principal Research Scientist
CSIRO
Tasmanian Food Research Unit
Hobart, Tasmania, Australia

T. B. S. Prakasam, Ph.D.
Research and Development Laboratories
Metropolitan Sanitary District
Cicero, Illinois

D. A. Ratkowsky, Ph.D.
Principal Research Scientist
Division of Mathematics and Statistics
CSIRO
Battery Point, Tasmania, Australia

Dale Sanders, Ph.D.
Department of Biology
University of York
York, England

Teixeira de Mattos, Ph.D.
Lecturer
Laboratory of Microbiology
University of Amsterdam
Amsterdam, The Netherlands

Karel van Dam, Ph.D.
Professor
Department of Biochemistry
B.C.P. Jansen Institute
University of Amsterdam
Amsterdam, The Netherlands

Hans V. Westerhoff, Ph.D.
Visiting Scientist
Molecular Biology Laboratory
National Institutes of Health
Bethesda, Maryland

CONTRIBUTORS

Volume II

James D. Bryers, Ph.D.
Professor
Center for Biochemical Engineering
Duke University
Durham, North Carolina

Antonio Casolari
Libero Docente
Department of Microbiology
Stazione Sperimentale
Parma, Italy

Raymond Leblanc, Ph.D.
Professor
Department of Mathematics and Computer Sciences
Université du Québec à Trois-Rivières
Trois-Rivières, Québec, Canada

Gerald M. Lefebvre, Ph.D.
Professor
Department of Physics
Université du Québec à Trois-Rivières
Trois-Rivières, Québec, Canada

Gerald Rosen, Ph.D.
M. R. Wehr Professor
Department of Physics and Atmospheric Science
Drexel University
Philadelphia, Pennsylvania

Paul R. Rutter, Ph.D.
Senior Chemist
Minerals Processing Branch
British Petroleum Research Center
Middlesex, England

Brian Vincent, D.Sc., Ph.D.
Reader in Physical Chemistry
Department of Physical Chemistry
University of Bristol
Bristol, England

TABLE OF CONTENTS

Volume I

Volume II

Chapter 7

MICROBIAL DEATH

Antonio Casolari

TABLE OF CONTENTS

I. INTRODUCTION

Microbial inactivation studies seem to be dominated by the exponential-single hit and multitarget theories, the former being applied mostly to heat inactivation kinetics and the latter to microbial inactivation by radiations. However, both theories are unable to satisfactorily explain some relevant experimental results, although they are valid enough to be of some predictive value in sterilization technology. Several satellite theories of lesser relevance do not add significantly to the understanding of inactivation phenomena.

In this chapter an attempt is made to avoid interpreting experimental data in too simplistic a fashion, and a general theory is presented which provides a unifying description of microbial inactivation kinetics.

A. Microorganisms and Lethal Agents

Microorganisms can be regarded as elementary biological particles able to undertake functional relationships within aqueous surroundings, whether provided with levels of autonomous functional organization (bacteria, yeasts, molds, etc.) or not (viruses). Characteristically, the primary function of microbial interaction with the environment is the production of progeny. Hence, the single practical criterion of death of microorganisms is the failure to reproduce in suitable environmental conditions.[1] Physical and chemical agents affecting microbial activities to such an extent as to deprive microbial particles of the expected reproductive capacity can be regarded as lethal agents.

Heat, ionizing, and UV radiations are the most relevant physical lethal agents. Ultrasonic frequencies, pressure, surface tension, etc. are mostly employed as cell disrupting agents in studying subcellular components. Chemical lethal agents compose a wide range of compounds employed in the microbial inactivation process called disinfection. The most important are oxygen, hydrogen peroxide, halogens, acids, alkalis, phenol, ethylene oxide, formaldehyde, and glutaraldehyde.

B. Shape of Survivor Curves

Survivor curves are usually described by plotting the logarithm of the number of microorganisms surviving against the size of treatment (time, dose of radiation, concentration of the chemical lethal agent). The semilogarithmic plot of survivor curves may have various shapes such as convex, sigmoid, concave, or linear (Figure 1). Different shapes can be obtained: (1) under identical experimental conditions with different microorganisms or with the same organism in a different physiological condition (i.e., vegetative cell or spore, the former being in the lag or log phase, the latter being activated or dormant, etc.); (2) using the same population of organisms while changing the destructive potential of the applied lethal agent (viz., changing the treatment temperature or the concentration of the chemical lethal agent); (3) using the same population of organisms while changing the environment in which the microorganisms are suspended or the culture medium employed to detect surviving fractions, etc.

Radiation inactivation studies show the typical convex survivor curve (often erroneously called sigmoid) characterized by a more or less extended lag in inactivation at lower doses (shoulder), followed by a nearly exponential decay phase (Figure 1(A)). Such a shape has also been found in heat inactivation experiments, although less frequently, and in microbial inactivation by chemical compounds.

Often disregarded, though not uncommon, is the true sigmoid shape characterized by a more or less pronounced tail occurring after initial shoulder and exponential decay phases (Figure 1(B)). Concave survivor curves characterized by a continuously decreasing death rate (CDDR) with increasing treatment time or size, and often described as bi- or multiphasic, can be produced by many physical and chemical lethal agents.

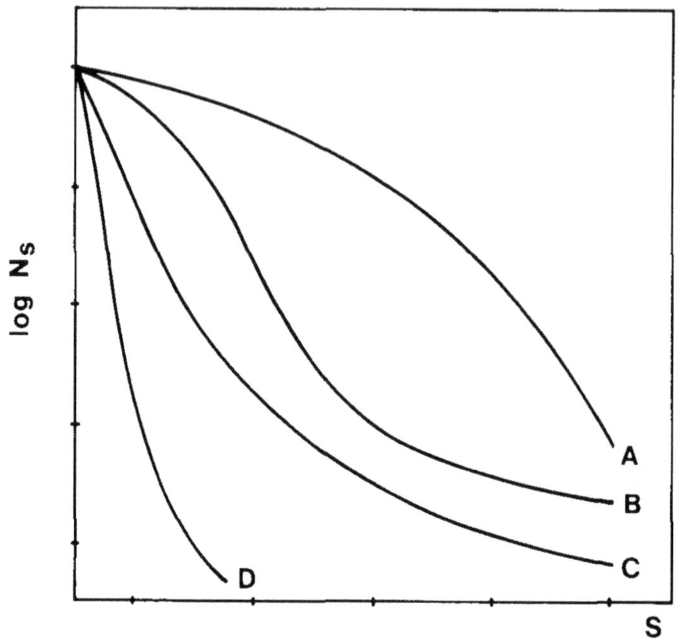

FIGURE 1. Shape of survivor curves of microorganisms treated with lethal agents: convex (A), sigmoid (B), and concave or continuously decreasing death rate curve (C and D). N_s is the concentration of survivors after the treatment size S.

The complexity of the situation was recognized about 70 years ago.[2-4] Nevertheless, most of the early authors attempted to describe semilog survivor curves by a straight line (exponential inactivation) regardless of shape, claiming that the direct utilization of the data would have been otherwise quite difficult. The most relevant practical application of the exponential simplification has been in heat sterilization technology. Reports of Bigelow and Esty[5] and Esty and Meyer[6] on the destruction rate of bacteria subjected to moist heat aided greatly in strengthening the belief in exponential inactivation, although counter to experimental evidence from their own results. Fundamental books on heat sterilization technology[7,8] perpetuated the belief in this exponential relationship, particularly the book by Stumbo,[8] who tried to explain any deviation from exponential decay with a series of conjectures, rather than with experimental evidence. Meanwhile, Pflug and Schmidt[9] claimed that "the majority of survivor curves are not straight lines on semilog plot", but did not propose an acceptable alternative theory.

II. SINGLE HIT THEORY

According to the single hit theory, the death of microorganisms results from the inactivation of a single molecule or site per cell; the death rate is expected to be proportional to the number of organisms remaining alive and follows first-order kinetics.

A. Heat Inactivation Kinetics

Plotting logarithm of surviving cell concentration, N_t, (N_t/g, N_t/mℓ, N_t/cm^2) against the time, t, of treatment at temperature T, a linear relationship is expected to occur:

$$\mathrm{Log}_{10} N_t = \mathrm{Log}_{10} N_o - k't \quad (1)$$

where N_o is the concentration of the untreated population and k' is the inactivation rate constant. This type of plot is regarded as very convenient if the resistance of different microorganisms is to be compared. In fact, a parameter called the decimal reduction time is easily obtained from the regression coefficient k' and is denoted by D_{90}, D_{10}, or D_T:

$$D_T = 1/k' \tag{2}$$

The use of D_{90} or D_{10} refers to the fact that after a treatment time $t = 1/k'$, 90% of the microbial population is destroyed or, alternatively, 10% of the population survives. In heat inactivation studies D_T is preferred, where T is the treatment temperature.

In exponential form, Equation 1 may be written:

$$N_t = N_o\ 10^{-k't} \tag{3}$$

so that after a treatment time $t = t_1$ Equation 3 becomes:

$$N_1 = N_o\ 10^{-k't_1} \tag{4}$$

and after treatment time $t = t_2$:

$$N_2 = N_o\ 10^{-k't_2} \tag{5}$$

Solving for k':

$$N_2/N_1 = 10^{-k'(t_2 - t_1)} \tag{6}$$

which in logarithmic form yields:

$$\mathrm{Log}_{10}(N_2/N_1) = -k'(t_2 - t_1) \tag{7}$$

so that:

$$-k' = (\mathrm{Log}_{10}N_2 - \mathrm{Log}_{10}N_1)/(t_2 - t_1) \tag{8}$$

and

$$k' = (\mathrm{Log}_{10}N_1 - \mathrm{Log}_{10}N_2)/(t_2 - t_1) \tag{9}$$

From Equation 2:

$$D_T = 1/k' = (t_2 - t_1)/(\mathrm{Log}_{10}N_1 - \mathrm{Log}_{10}N_2) \tag{10}$$

and, as expected, if the treatment time is increased from t_1 to t_2, 90% of the population is destroyed, $\mathrm{Log}_{10}\ N_1 - \mathrm{Log}_{10}\ N_2$ equals 1 and the decimal reduction time at temperature T is

$$D_T = t_2 - t_1 \tag{11}$$

Each type of microorganism (virus, bacterial vegetative cell or spore, yeast vegetative cell or spore, fungal cell or spore), as well as each species and strain of the same group of organisms, has its own resistance at a particular temperature T under defined environmental conditions, that is, its own D_T. On changing the environment, the microbial resistance changes accordingly.

The most relevant physicochemical factors affecting heat resistance are the water content in the environment, pH, and temperature. Changing the solute concentration in the medium also changes the osmotic pressure or, referring to the usual parameter employed in food microbiology, the water activity a_w ($a_w = p/p_o$, where p and p_o are the water pressures of the medium and of pure water, respectively, in isothermal and isobaric conditions). The water activity is described by Raoult's law, so that:

$$a_w = f(n_w/(n_w + n_s)) \tag{12}$$

where n_w and n_s are the concentrations of water and solute, respectively. It follows that by increasing the solute concentration, the a_w of the environment decreases. As a_w decreases, the thermal resistance of microorganisms increases. The D_T value in dry conditions may be more than 10^3 times higher than that in moist (high a_w) conditions.[9] An exact relationship between D_T and a_w has not been developed.

Thermal resistance is usually higher at neutral pH and it decreases when the pH is increased or decreased. A defined relationship between D_T and pH is not known, though a tenfold change in D_T is often observed for each 2 pH units.

With respect to chemical reactions, a defined relationship between D_T and temperature can be established. Plotting Log D against temperature, a linear relationship is usually obtained:

$$\mathrm{Log}_{10} D_T = \mathrm{Log}_{10} U - bT \tag{13}$$

where U is a proportionality constant and b is the rate at which D changes with temperature. Equation 13 may be written:

$$D_T = U\ 10^{-bT} \tag{14}$$

The above equations yield the parameter z:

$$z = 1/b \tag{15}$$

which is related to the temperature coefficient Q_{10} by:

$$z = 10/\mathrm{Log}_{10} Q_{10} \tag{16}$$

where

$$Q_{10} = k_{T+10^o}/k_T \tag{17}$$

z is the number of degrees required to achieve a tenfold change in D_T. It follows that:

$$0.1\ D_{T-z} = D_T = 10\ D_{T+z} \tag{18}$$

If we let D_1 and D_2 be the D_T values at the temperature $T_1 < T_2$, respectively, it follows that:

$$D_1 = U\ 10^{-bT_1} \tag{19}$$

and

$$D_2 = U\ 10^{-bT_2} \qquad (20)$$

so that, solving for b:

$$D_2/D_1 = 10^{-b(T_2 - T_1)} \qquad (21)$$

and in logarithmic form:

$$\mathrm{Log}_{10}(D_2/D_1) = -b(T_2 - T_1) \qquad (22)$$

and then:

$$b = (\mathrm{Log}_{10}\ D_1 - \mathrm{Log}_{10}\ D_2)/(T_2 - T_1) \qquad (23)$$

and

$$z = 1/b = (T_2 - T_1)/(\mathrm{Log}_{10}\ D_1 - \mathrm{Log}_{10}\ D_2) \qquad (24)$$

As expected, when $D_1 = 10\ D_2$, the z value will be equal to $T_2 - T_1$.

Knowledge of the D_T and z values for microorganisms in given environmental conditions is very useful in practice, when heat sterilization times at an unknown temperature T_u must be determined when D_{Tu} is not known. In such a case,

$$T_u - T = nz \qquad (25)$$

$$D_{Tu} = D_T\ 10^{-(T_u - T)/z} \qquad (26)$$

$$D_{Tu} = D_T\ 10^{-n} \qquad (27)$$

$$D_{(T+nz)} = D_T/10^n \qquad (28)$$

and so:

$$t_{(T+nz)} = t_T/10^n \qquad (29)$$

where $t_{(T+nz)}$ is the treatment time at the temperature $T + nz$ equivalent to the time t_T at the reference temperature T.

Sterilization cycles are usually based on a minimum time required to obtain 12 decimal reductions of a particular microorganism (*Clostridium botulinum* spores in food sterilization technology, for instance). Usually Equation 29 is employed, although using the symbolism:

$$\tau_{(T+nz)} = \tau_T/10^n \qquad (30)$$

where τ is the number of minutes required to obtain the expected number, i, of decimal reductions ($\tau = i\ D$). In heat sterilization technology, at the reference temperature T = 121.1°C the $\tau_{121.1}$ value is F_o and $z = 10$°C, so that Equation 30 becomes:

$$\tau_{121.1}/\tau_{(T+n\ 10)} = 10^n \qquad (31)$$

or in a better known form,

$$F/\tau = 10^{(T_{(T+n\ 10)} - 121.1)/10} \tag{32}$$

Equation 32 is currently employed to evaluate the efficiency of sterilization cycles.[7,8]

By analogy with chemical kinetics, the exponential heat inactivation curve can be described by the relationship:

$$-dN_t/dt = kN_t \tag{33}$$

yielding, by integration:

$$\mathrm{Log}(N_t/N_o) = -kt \tag{34}$$

and:

$$N_t/N_o = \exp(-kt) \tag{35}$$

It follows that D_T as defined by Equation 10 is

$$D_T = 2.303/k \tag{36}$$

The analogy can be extended to the relationship between k and temperature. According to chemical kinetics:

$$k = A\ \exp(-E_a/RT) \tag{37}$$

where A is a frequency factor, E_a is the Arrhenius activation energy, R is the gas constant (1.987 cal/mol), and T is the absolute temperature (degrees Kelvin).

From Equation 37:

$$k_1 = A\ \exp(-E_a/RT_1) \tag{38}$$

and

$$k_2 = A\ \exp(-E_a/RT_2) \tag{39}$$

so that:

$$k_1/k_2 = \exp(-E_a(T_1^{-1} - T_2^{-1})/R) \tag{40}$$

From Equation 40,

$$\mathrm{Log}(k_1/k_2) = (-E_a/R)(T_2 - T_1)/(T_2T_1) \tag{41}$$

and since $k = 2.303/D_T$, from Equation 36,

$$\mathrm{Log}(D_2/D_1) = (-E_a/R)(T_2 - T_1)/(T_2T_1) \tag{42}$$

and

$$\mathrm{Log}_{10}(D_2/D_1) = (-E_a/2.303\ R)(T_2 - T_1)/(T_2T_1) \tag{43}$$

Taking into account Equation 24,

$$z = 2.303\ R\ T_1T_2/E_a \tag{44}$$

It follows that the z value cannot be regarded as being constant, but varies with temperature. It follows that a z value of 5°C obtained between 60 and 65°C increases to 5.23 between 65 and 75°C, and to 5.53 between 75 and 85°C. Analogously, a z value of 10°C obtained in the temperature range 100 to 110°C will reach the value of 11.09 in the range 120 to 130°C.

According to Equation 44, the Arrhenius relationship requires that z increases as the temperature increases. The constancy of z values obtained by plotting $Log_{10}\ D_T$ against temperature is difficult to ascertain in practice, as the evaluation of D is not sufficiently accurate. In some instances, a trend toward an increase of z with increasing temperature has been shown.[10-14] Following chemical kinetics, the Arrhenius activation energy E_a may be obtained using Equation 37, plotting Log k against T^{-1}. Alternatively, E_a can be obtained from:

$$E_a = a\ Log\ Q_{10} \tag{45}$$

or

$$E_a = c/z \tag{46}$$

where $a = 0.1\ R\ T_1T_2$ and $c = 2.303\ R\ T_1T_2$, since, according to Equation 43:

$$E_a = 2.303\ Log_{10}(D_1/D_2)\ R\ T_1T_2/(T_2 - T_1) \tag{47}$$

Nevertheless, z is not constant, according to Equation 44, so E_a values obtained from Equation 46 differ from those from the Arrhenius equation (Equation 37) by about 1 kcal/mol (4.18 kJ/mol). Figure 2 shows the expected value of E_a as a function of Q_{10} and z in temperature ranges resulting in the thermal destruction of microorganisms.

The standard energy of activation, ΔG, the standard enthalpy of activation, ΔH, and the standard entropy of activation, ΔS, are sometimes reported for the thermal death of microorganisms. According to Eyring's relationship:

$$k = k^{*}(KT/h)\ k_c \tag{48}$$

where K is the reaction rate constant observed at the absolute temperature T, k^{*} is the transmission coefficient (usually considered $\simeq 1$), K is Boltzmann's constant ($1.38 \cdot 10^{-16}$ erg/degree), h is Planck's constant ($6.624 \cdot 10^{-27}$ erg/sec), and k_c is the equilibrium constant; k_c may be expressed in terms of the standard energy of activation:

$$k_c = exp(-\Delta G/RT) \tag{49}$$

and since:

$$\Delta G = \Delta H - T\Delta\ S \tag{50}$$

Equation 48 becomes:

$$k = (KT/h)\ exp(\Delta S/R)\ exp(-\Delta H/RT) \tag{51}$$

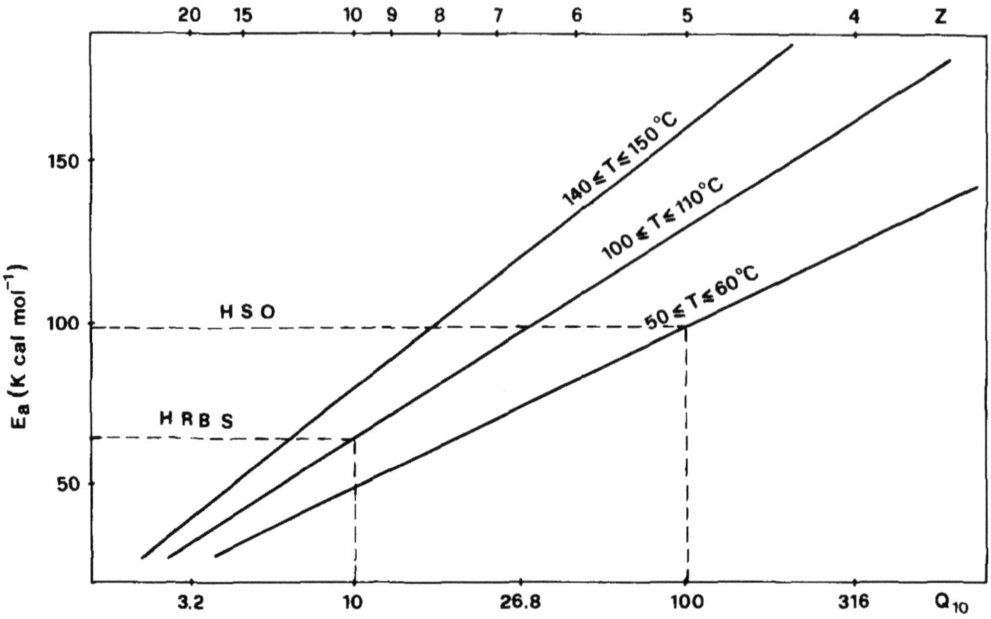

FIGURE 2. Relationship between Q_{10}, corresponding z values, and Arrhenius activation energy E_a, predicted by thermodynamic treatment of first-order inactivation kinetics of microorganisms subjected to low temperatures (50 to 60°C) (heat sensitive organisms, HSO, with an average Q_{10} = 100), or to high temperatures (≥100°C; heat resistant bacterial spores, HRBS, with an average Q_{10} = 10).

Equation 51 is very similar to the Arrhenius relationship, taking into account that:

$$\exp(-\Delta H/RT) \simeq \exp(-E_a/RT) \quad (52)$$

when $E_a \gg RT$, since:

$$\Delta H = E_a - RT \quad (53)$$

and from Equations 37 and 51:

$$A = (KT/h)\exp(\Delta S/R) \quad (54)$$

Because, in microbial heat inactivation kinetics RT is more than 100 times lower than E_a, Equation 51 can be written:

$$k = (KT/h)\exp(\Delta S/R)\exp(-E_a/RT) \quad (55)$$

after which:

$$\Delta S = R(\text{Log } k - \text{Log}(KT/h) + E_a/RT) \quad (56)$$

or, taking into account that $K = \underline{R}/N$, where N is Avogadro's number and $\underline{R} = 8.32 \cdot 10^7$ erg/mol/degree:

$$\Delta S = R(\text{Log } k - \text{Log}(\underline{R}\ T/Nh) + E_a/RT) \quad (57)$$

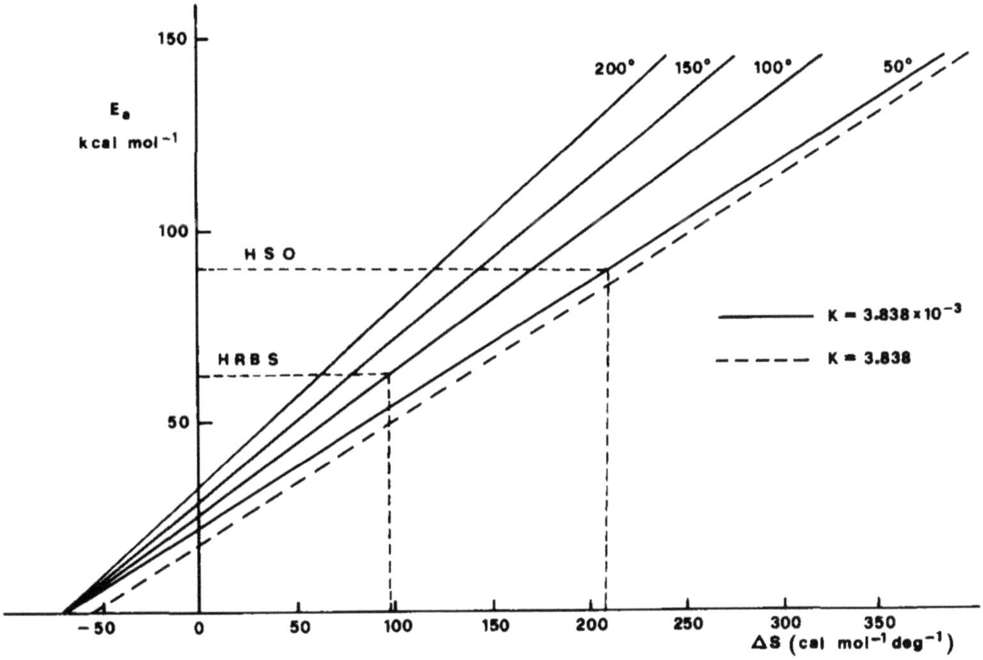

FIGURE 3. Relationship between Arrhenius activation energy E_a (i.e., the activation enthalpy minus RT) and activation entropy (ΔS), predicted by thermodynamic treatment of first-order inactivation kinetics of microorganisms subjected to 50°C (heat sensitive organisms, HSO) or to higher temperatures (heat resistant bacterial spores, HRBS).

As shown in Figure 3, there is a linear relationship between E_a or ΔH and ΔS at any given temperature. It follows that Equation 57 (or Equation 56) can be reduced to:

$$\Delta S = (E_a/T) + B \tag{58}$$

where B equals R(Log k + Log ($\underline{R}$T/Nh)) or R(Log k + Log (KT/h)). Equation 58 is often called the "isokinetic relationship" or the "compensation law", since an increase in the activation energy or enthalpy is exactly compensated by an increase in entropy.[15-17] Nevertheless, when activation energy and activation entropy values are obtained by a series of computations based on Equations 37 to 58, the compensation law is not such a mysterious relationship as it is sometimes regarded, but follows directly from the premises (Equation 48).[18,19]

At this point, the analogy between microbial heat inactivation kinetics and chemical reaction kinetics should be examined closely.

1. Thermodynamic Inconsistencies

Figure 4 shows the frequency distribution of 231 z values collected from the literature. As can be seen, the spread is very large. After grouping collected data into 5°C classes the most frequent z value obtained for viruses and microbial vegetative cells is 5°C (Q_{10} = 100) and for bacterial spores is 10°C (Q_{10} = 10). It must be remembered that viruses and vegetative microbial cells are usually 10^2 to 10^8 times less heat resistant than bacterial spores. Vegetative cells are destroyed at temperatures ranging from 50 to 60°C at a rate k ≃ 3.838 · 10^{-3} (D_T ≃ 10 min). Bacterial spores are destroyed at a similar rate at temperatures ranging from 100 to 120°C.

As shown in Figure 2, the activation energy obtained for less resistant vegetative cells

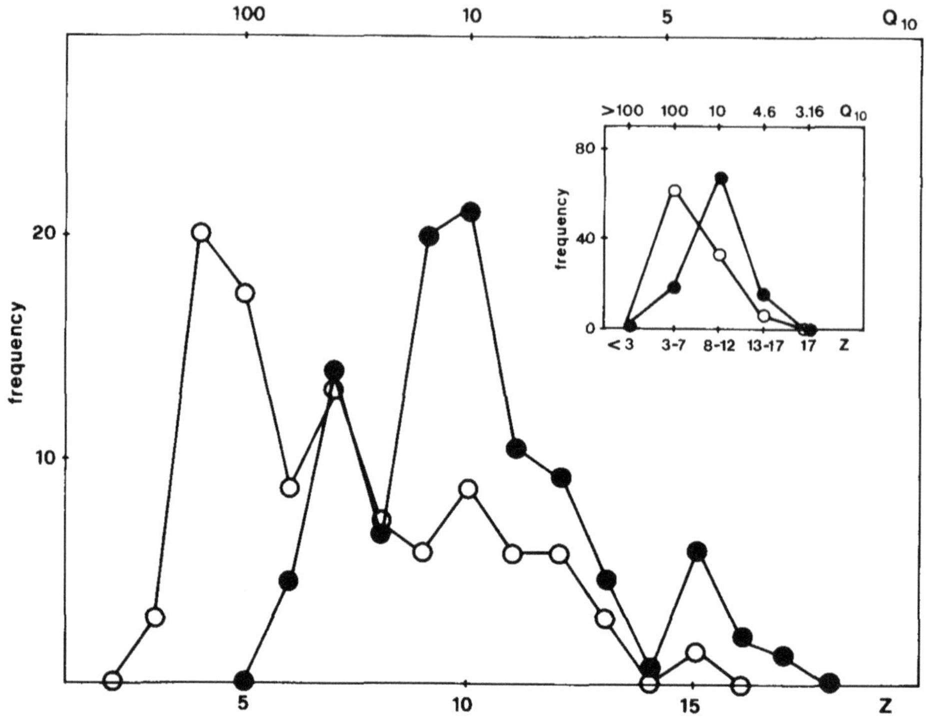

FIGURE 4. Frequency distribution curves of 231 z values collected from the literature, together with corresponding Q_{10} values, for viruses, vegetative bacteria, yeasts, and molds (○) and bacterial spores (●).

(mean z = 5°C) is much higher than the E_a value of more resistant bacterial spores (mean z = 10°C), the former being about 90 kcal/mol (380 kJ/mol) and the latter about 60 kcal/mol (250 kJ/mol).

Disregarding the parameter A in Equation 37, reaction rates of analogous reactions are expected to be inversely related to the activation energy value required for the occurrence of single reactions. Contrariwise, the heat inactivation of bacterial spores has a lower activation energy, while occurring at a rate very much slower than that of heat-sensitive viruses or vegetative microbial cells. Consequently, it can be argued, the A value in the Arrhenius equation cannot be disregarded and Eyring's relationship must be taken into account.

As previously shown, the so-called compensation law requires that high E_a (or ΔH) values necessarily follow high ΔS values. In fact, as shown in Figure 3, ΔS for vegetative cells is about 210 cal/mol/degree (880 J/mol/degree) and ΔS for bacterial spores is about 90 cal/mol/degree (380 J/mol/degree). Accordingly, higher activation energies required to inactivate vegetative cells would be justified on the basis of the greater activation entropy involved in the process. Disregarding uncertainties associated with the rigid application of the compensation law, it seems hard to envisage a greater ΔS for the heat inactivation of vegetative cells rather than of spores, since it is reasonable to assume that the entropy level of the former is higher than that of the more orderly assembled molecular structures of bacterial spores; however, this may not be the case.

Nevertheless, the reservations regarding what results merely from a thermodynamic treatment of exponential heat inactivation kinetics are further strengthened by the very high values of the ensuing E_a or ΔH. Using the Maxwell-Boltzmann's law for the distribution of velocities among molecules:

$$n_E/n_o = 2\pi^{-\frac{1}{2}} (E_a/RT)^{\frac{1}{2}} \exp(-E_a/RT) \quad (59)$$

where n_E and n_o are the number of molecules having energy E_a and the total number of molecules, respectively; or in the simplified form (i.e., on the basis of two axes):

$$n_E/n_o = \exp(-E_a/RT) \quad (60)$$

For values of $2\pi^{-1/2}(E_a/RT)^{1/2}$ ranging between about 5 and 15 at temperatures $\geq$ 273.15K and E_a values lower than about 100 kcal/mol, it can be computed that the probability of finding E_a values as high as those obtained for vegetative cells exposed to lethal temperatures is so low as to be unreasonable. The probability $P(E) = n_E/n_o$ of occurrence of molecules having E_a = 90 kcal/mol at 60°C is expected to be, according to:

$$n_{90,000}/6 \cdot 10^{23} = \exp(-90{,}000/((60 + 273.15)\ 1.987)) = 8.99 \cdot 10^{-60}$$

which means that a single molecule carrying more than 90 kcal could occur, at 60°C, in about 10^{36} mol of a substance (i.e., a mass of hydrogen about 10^3 times that of the sun). A mole of cells weighs about 10^{11} g; therefore, we would find a single molecule carrying the above energy in 10^{47} g of cells, a weight about 10^3 times that predicted for our galaxy. Summing the number of intermonomer chemical bonds of about 10^6 protein molecules (average mol wt, $2 \cdot 10^4$) and of DNA (mol wt about 10^9) occurring in a microbial cell, we obtain about 10^{10} chemical bonds per cell. This implies that we would expect to find a single one of the above 10^{10} bonds with about 100 kcal in a mass of cells approximately 10^4 times that of the sun. The above results seem to suggest that E_a values as high as those obtained by the thermodynamic treatment usually applied to microbial heat inactivation kinetics can be regarded to be of doubtful meaning.

a. Maximum Allowable Q_{10}

Q_{10} values obtained in microbial heat inactivation studies are usually much higher than those expected to occur in chemical kinetics. The maximum Q_{10} (MQ_{10}), the value above which thermodynamic treatment of reaction rate loses any meaning, can be evaluated. According to Equations 45 and 60, and letting P(E) be the probability of occurrence of molecules carrying more than E_a energy, it follows that:

$$P(E) = \exp(-a \operatorname{Log} Q_{10}/RT) \quad (61)$$

The minimum value of the probability (mP(E)) required for a reaction to occur at the temperature $T_2 = T_1 + 10°$ can be regarded to be 1, that is, at least one n_E molecule can be expected to be present in the total amount of available reactants, or one n_E per total number of available molecules; so that:

$$mP(E) = 1/iN \quad (62)$$

where i is the number of moles of available reagents and N is the Avogadro's number. If the above probability is higher at T_2 than at T_1, Equation 61 can be written:

$$mP(E) = \exp(-a \operatorname{Log} Q_{10}/RT_2) \quad (63)$$

so that combining Equations 62 and 63:

$$(iN)^{-1} = \exp(-a \operatorname{Log} Q_{10}/RT_2) \quad (64)$$

Table 1
MAXIMUM ALLOWABLE Q_{10} AND MINIMUM ALLOWABLE Z EXPECTED FROM THERMODYNAMIC TREATMENT OF FIRST-ORDER INACTIVATION KINETICS OF 1 g (A) OR 1 kg (B) OF MICROBIAL PARTICLES HEATED AT THE TEMPERATURE $T_2 = T_1 + 10°C$

		Viruses $(3 \cdot 10^{15})$[a]		Bacteria $(6 \cdot 10^{11})$[a]				Yeasts $(8.6 \cdot 10^{9})$[a]	
				Vegetative		Spores			
T_2		$Q^{\cdot}_{10}$	$z^{\cdot}$	$Q^{\cdot}_{10}$	$z^{\cdot}$	$Q^{\cdot}_{10}$	$z^{\cdot}$	$Q^{\cdot}_{10}$	$z^{\cdot}$
60	A	2.91	21.5	2.26	28.3	b		1.99	33.5
	B	3.59	18.0	2.78	22.5	b		2.44	25.8
70	A	2.83	22.2	2.20	29.1	b		1.95	34.5
	B	3.46	18.6	2.70	23.2	b		2.38	26.5
100	A	c		c		2.07	31.7	c	
	B	c		c		2.49	25.3	c	
120	A	c		c		1.99	33.4	c	
	B	c		c		2.38	26.6	c	

Note: T_2 = °C; $Q^{\cdot}_{10}$ = maximum allowable Q_{10}; $z^{\cdot}$ = minimum allowable z.

[a] Total number of particles/g.
[b] Temperature too low for destruction of bacterial spores.
[c] Temperature too high for heat-sensitive microbial particles.

As $a = 0.1\ R\ T_1T_2$, as previously shown, and rearranging:

$$MQ_{10} = (iN)^{10/T_1} \tag{65}$$

it follows that, letting the highest conceivable value of $i = 10^3$, at $T_1 = 50$ or 100°C the maximum allowable Q_{10} would be 6.74 or 5.22, respectively.

A Q_{10} of 2 to 3, as is usually found in chemical kinetics, is expected to be obtained at about 100°C in an amount of solution containing about 10^{-12} to 10^{-6} mol of reagents, that is, about 10^{11} or 10^{18} molecules, respectively ($i = (MQ_{10})^{T_1/10} \cdot N^{-1}$).

Table 1 shows the MQ_{10} values expected from a thermodynamic treatment of microbial heat inactivation, according to Equation 65, for two amounts of cells (1 g and 1 kg), exceedingly higher than those employed in practice. By comparing values reported in Table 1 with those in Figure 4, it can be seen that all Q_{10} values obtained by treating microbial heat inactivation as a first-order reaction, can be regarded as unreasonably high.

b. Minimum Allowable z

According to Equation 17, if Q_{10} increases, z decreases. The minimum allowable z value, z_m, under which P(E) must be regarded as being unreasonably low, can also be obtained. From Equations 17 and 65:

$$(iN)^{10/T_1} = 10^{10/z_m} \tag{66}$$

Taking logarithms and rearranging:

$$z_m = T_1/\mathrm{Log}_{10}(iN) \tag{67}$$

Microbial heat inactivation experiments are usually done employing less than 10^{12} particles per liter of medium. Therefore, while taking into account the maximum allowable concentration of cells that could be employed in practice, the minimum allowable z values following Equation 67 are two to six times higher than those found in the literature (Table 1). We could assume *ab absurdo* that a microorganism is inactivated if a single chemical bond in the cell is broken, so that the quantity iN in Equation 67 could be regarded to be about 10^{25}; the resulting z_m will be 13.33, 14.13, 14.93, or 15.73 at T_1 = 60, 80, 100, or 120°C, respectively. As a consequence, all z values reported in the literature (Figure 4) can be regarded as unreasonably low.

c. The Highest Allowable Energy Value

It would be interesting to know the highest reasonable E_a value. To this end, Equation 63 can be written:

$$mP(E) = \exp(-ME_a/RT_2) \tag{68}$$

where ME_a is the maximum value of the energy of activation at the temperature $T_2 = T_1 + 10°C$, and thus the E_a value above which P(E) is too low to be acceptable. From Equation 68:

$$ME_a = R\,T_2(-\text{Log}\ mP(E)) \tag{69}$$

and then:

$$ME_a = R\,T_2(-\text{Log}(1/iN))$$

$$= R\,T_2\ \text{Log}(iN) \tag{70}$$

It follows that at T_2 = 50 or 100°C the maximum allowable E_a value is expected to be 39,591 or 45,716 cal/mol (165,708 or 191,348 J/mol).

Table 2 shows expected values of ME_a according to Equation 70 for first-order kinetics of microbial heat inactivation. As can be seen, all E_a values reported in the literature are unreasonably high.

Perhaps as a provisional conclusion we may say that the thermodynamic treatment of reaction kinetics on an exponential basis is not readily applicable to the heat inactivation of microorganisms. It can be rather useful to retrace our steps in order to answer, first, the most relevant question: could microbial heat inactivation kinetics be regarded as an exponential phenomenon? Several observations seem to suggest that this is not always the case. The question will be examined later.

2. *Mechanisms and Models*

The greatest insight into heat inactivation mechanisms was obtained by Ball and Olson,[7] as presented in the book *Sterilization in Food Technology,* still regarded as the bible of the field. Through an analysis of the concepts of heat and temperature on a macroscopic and microscopic scale, they pointed out that "whenever Brownian movement is observed, our macroscopic concepts of temperature and heat transfer break down and must be replaced by energy considerations involving molecules in the discrete, and not in the statistical sense" (see also Tischer and Hurwicz[20]). Soon after " . . . it is not something within the cell (such as temperature) which is the cause of death. The cause must be outside the cell. It must be in the medium."[7] Accordingly, Ball and Olson suggested a mechanism of microbial death brought about by " . . . one or more molecules in the surrounding medium", having "the greater mean velocity . . . according to the velocity distribution curve". Nevertheless,

Table 2
MAXIMUM ALLOWABLE VALUES OF THE ACTIVATION ENERGY E_a EXPECTED FROM THERMODYNAMIC TREATMENT OF FIRST-ORDER INACTIVATION KINETICS OF 1 g (A) OR 1 kg (B) OF MICROBIAL PARTICLES HEATED AT THE TEMPERATURE $T_2 = T_1 + 10°C$

			Bacteria		
T_2 (°C)		Viruses $(3 \cdot 10^{15})$[a]	Vegetative $(6 \cdot 10^{11})$[a]	Spores	Yeasts $(8.6 \cdot 10^{9})$[a]
			E_a (cal/mol)		
Sensitive organisms					
70	A	24298	18491		15597
	B	29008	23201		20307
80	A	25007	19030		16051
	B	29854	23877		20898
Resistant organisms					
110	A			20647	
	B			25906	
130	A			21724	
	B			27258	

[a] Total number of particles/g.

they did not develop a mathematical model for microbial death.

Charm[21] presented a model referring to the discrete nature of molecules involved in heat inactivation mechanisms, although diverging somewhat from the work of Ball and Olson. Charm's model leads to exponential inactivation, but some uncertainties cast doubts on its reliability. Charm[21] regards the microbial cell as " . . . composed of a number of sensitive volumes surrounded by a number N of water molecules; when the water molecule, in contact with a sensitive volume, is able to impart sufficient energy to the sensitive volume inside the cell, it is thought to cause inactivation of the cell". The model yields the equation:

$$\log(N_t/N_o) = -S\ t \exp(-E/RT) \qquad (71)$$

where S (min^{-1}) is the frequency with which water molecules exchange the energy, E, with the sensitive volume.

Some inconsistencies arising from the model can be pointed out: (1) Charm defined E as energy/cell, while energy/mol was considered, since R = 1.987 cal/mol/degree was seemingly taken throughout for computation purposes; (2) reported E values seem to be too high (for *C. botulinum* spores heated in distilled water, for instance, E = 73,500 Btu/spore, equivalent to about $1.85 \cdot 10^7$ cal or $7.35 \cdot 10^7$ J); (3) the reported values of the frequency factor S are much too high to be reasonable ($S \geq 10^{20}$), being equal to or greater than the frequency of X-rays produced by a potential difference of about $1.7 \cdot 10^5$ V. In addition, Charm's model contains the same inconsistencies pointed out previously regarding the thermodynamic treatment of exponential kinetics (i.e., very high energy values, energy values greater for less than for more resistant organisms, etc.) and substantially, it leaves inactivation kinetics still unresolved.

B. Radiation Inactivation

At present, studies of microbial inactivation by UV radiation are usually carried out using low pressure mercury lamps which emit about 95% of their light at a wavelength very close to 253 nm. The plot of the logarithm of surviving organisms against UV dose yields, very frequently, nonlinear curves, with a pronounced tendency to tail off.[22-25] A quantum of UV radiation has an energy level of about 5 eV, insufficient to eject electrons from atoms or molecules, and able to cause only excitation processes. High energy radiations (γ and X-rays, α, β, and neutron particles) have, characteristically, the property of causing ionization in the absorbing material; many atoms and molecules are excited and converted to free radicals.[26] The density of ionization and excitation events produced along the path of radiation depend upon the photon energy and the type of absorbing material, and is usually expressed as a linear energy transfer (keV/μm). Linear energy transfer (LET) increases with the square of the charge carried by the particle and decreases as its speed increases.[26] The relative biological effectiveness usually increases as the LET of the employed radiation increases, but this is not always the case.[27,28]

A very limited number of experimental survival curves of irradiated microorganisms can be safely described by the law of exponential decay:

$$N_d/N_o = \exp(-a\ d) \tag{72}$$

where N_d is the number of microorganisms surviving the dose d of radiation and a is the rate constant of inactivation. The coefficient a in Equation 72 is expected to represent the probability of inactivation of microorganisms per unit dose (rad or gray) of radiation. Inactivation is expected to result from a single hit, that is, through excitations and ionizations occurring along the track of photons crossing the cell. Hence, the concept of "direct action" arises, according to which the target is a molecule or a narrow group of molecules within the cell, in which the primary event (i.e., an ionization) has to occur, or through which or near which a unit of ionizing radiation must pass.[29]

On the other hand, the "indirect action" theory assumes that the whole solution, in which microorganisms are suspended, is the true target, lethal events being produced by reactions of chemical active agents induced by radiation in water, outside, and/or inside the cell. In fact, a number of reactions do occur in irradiated water:

$$H_2O \rightarrow H_2O^+ + e^-$$
$$H_2O^+ \rightarrow OH^\cdot + H^+$$
$$e\ + H_2O \rightarrow OH^- + H^\cdot$$
$$2H^\cdot \rightarrow H_2$$
$$2OH^\cdot \rightarrow H_2O_2$$
$$H^\cdot + OH^\cdot \rightarrow H_2O$$
$$H^\cdot + O_2 \rightarrow {}^\cdot HO_2$$
$$2\ {}^\cdot HO_2 \rightarrow H_2O_2 + O_2$$

The most active and long-living radicals are regarded to be $OH^\cdot$. For the single-hit theory to be tenable, direct interaction between the radiation and a single sensitive target inside the cell must occur.[30] However, many environmental factors can modify the sensitivity of microorganisms to ionizing radiations such as oxygen, moisture, the presence of sulfydryl compounds and/or –SH reagents in the medium, previous treatments of the microorganisms, physiological age of the cells, etc.[31-36] Actually, exponential survival curves of irradiated microorganisms usually result from experiments carried out either using very sensitive organisms or are not prolonged enough to determine the shape of survivor curves at very low surviving fractions. Nevertheless, it is well known that survivor curves of irradiated microorganisms are not usually exponential, but are typically sigmoid.

C. Chemical Inactivation

The concept of "effective concentration" requires that a lethal chemical compound must possess the ability to reach, and to accumulate at, the site(s) of action, whether on the surface of, or within, the microbial cell. Thus, a dose-effect curve is expected to be characteristically sigmoid. At very low concentrations of the lethal compound the microbial inactivation rate will be very low and approaches zero; at higher doses, the death rate is expected to increase as the dose itself is increased; at the highest doses, the rate of the process would not increase further, since it would be sorption-limited. In spite of these expectations, Madsen and Nyman[2] and Chick[3] established a mathematical model for chemical disinfection on the basis of an analogy between microbial inactivation process and first-order reaction kinetics. This model has formed the basis of most subsequent investigations.

According to the model, the relationship between the surviving organisms and the contact time, t, with a given concentration of a lethal chemical compound is expected to be

$$N_t/N_o = \exp(-k_t\, t) \qquad (73)$$

where k_t is the rate constant. The inactivation rate constant k_t changes as a function of the concentration of the disinfectant. The relationship between the rate constant k_t and the concentration C of the lethal agent is

$$k_t = B \exp(k_c\, C) \qquad (74)$$

where k_c is the rate of change of inactivation rate per unit change in the concentration of the lethal compound. Usually Equation 75 is used instead of Equation 74:

$$k_t' = B'\, 10^{k_c'\, C} \qquad (75)$$

and $1/k_c'$ is the concentration required to achieve a tenfold change in the inactivation rate.

However, subjecting microbial populations to increasing concentrations of lethal compounds for a fixed contact time seldom produces survival curves described by a function of the type:

$$N_d/N_o = \exp(-k_d\, D) \qquad (76)$$

where D is the concentration of the lethal compound. Usually Equation 77 is used instead of Equation 76:

$$N_d/N_o = 10^{-k_d'\, D} \qquad (77)$$

and $1/k_d'$ equals the change in concentration required for the survival probability N_d/N_o to change ten times.

Reaction rate constants k_t, k_c, and k_d are expected to increase as temperature increases, following the Arrhenius law:

$$k_{t,c,d} = F \exp(-E_a/RT) \qquad (78)$$

where F is a frequency factor whose value is linked to the rate constant considered (k_t, k_c, or k_d).

Actually, a relationship of the above type is expected to occur in disinfection processes at temperatures lower than about 45 or 100°C for vegetative cells (heat sensitive) or bacterial spores (heat resistant), respectively. In a range of temperatures sufficiently high for heat inactivation, a relationship more complex than Equation 78 is expected to be more appropriate.

A strikingly limited number of Arrhenius plots is reported in the literature, notwithstanding the relevance of temperature coefficients of chemical lethal compounds to the practice of disinfection, especially with respect to the choice of suitable agents of disinfection. Q_{10} values for microbial inactivation by chemical compounds reported in the literature range between about 1.6 and 3.3 and have been obtained by treating bacterial spores with hydrogen peroxide, formaldehyde, glutaraldehyde, β-propiolactone, ethylene oxide, chlorine, and iodine, at temperatures ranging from −10 to 95°C.[37-42] Therefore, Arrhenius activation energies are expected between about 10 and 25 kcal/mol. Recently, Gelinas et al.[43] reported E_a values ranging from 0 to about 37 kcal/mol, so that Q_{10} fell between about 1 and 7.7, although, the method used by the authors to assay the sensitivity of vegetative bacterial cells cannot be regarded as being reliable. Higher E_a values are reported for *Escherichia coli* treated with phenol at temperatures from 30 to 42°C: 52 kcal/mol;[44] a $Q_{10} = 10$ was reported using peracetic acid.[45]

Nevertheless, a disinfection model is difficult to develop on the basis of these sorts of results, since several observations suggest that survivor curves may have different shapes (mostly convex, sigmoid, and concave, with a more or less pronounced tail), according to the experimental conditions employed.[37,40-42,45-49] A comprehensive theory of the disinfection process is, unfortunately, still lacking.

D. Theoretical Uncertainties

Characteristically, the single-hit theory is applied to heat inactivation kinetics. Heat sterilization technology is based on the tenet of exponential inactivation. If this order of death should be found to be invalid, the efficiency of the sterilization process becomes questionable.

As pointed out by several authors, first-order kinetics are expected to be produced by unimolecular reactions, without regard to the underlying reasons (pseudo-first-order reaction, for instance).

As far as microbial inactivation is concerned, a unimolecular reaction would comply with the single-hit or single-site theory, after which a single damage produced in the cell would unequivocally lead to the death of the cell, whether affecting an enzyme, the DNA, or a different molecule. The concept is quite limiting, since microorganisms subjected to a lethal agent are not unequivocally dead or alive, but they either die or recover, depending on the environmental conditions applied after the treatment. The phenomenon is well known to microbiologists and is called "sublethal injury or damage". Usually, microorganisms treated with lethal agents become more exacting about their environmental conditions than untreated organisms.[12,24,34,50-57] As a consequence, survivor curves obtained using a population subjected to a lethal treatment might differ depending on the experimental conditions applied after the treatment. As a rule, the extent of the damage becomes increasingly difficult to demonstrate as the intensity of the applied lethal condition increases. The sublethal injury phenomenon can hardly be reconciled with the single-hit theory. On the contrary, it seems to suggest that microbial death can be rather more satisfactorily envisioned as an end point of a gradual, damaging process.

III. TARGET THEORY

A. Inactivation by Radiation

The effects of ionizing radiation upon microorganisms have led to the development of some very interesting models, usually described under the name of "target theory".

The concept of microorganisms, or something(s) inside microorganisms, as targets being hit by photons or particles, reflects quite closely what is believed to occur in radiological phenomena. The basic assumption of the "multiple hit" theory was that a single target must be hit n times before the organism is destroyed.[58-60] Let a be the sensitive volume of the cell and h the average number of hits per unit volume of the microbial suspension; then, ah is the average number of hits within the sensitive volume a. If the hits occur independently and at random, the probability that f hits fall within the volume a is given by the Poisson distribution:

$$P(f) = \exp(-ah) \cdot (ah)^f/f! \tag{79}$$

If n is the number of hits required to inactivate a microorganism, then all cells receiving less than n hits will survive. The probability N_h/N_o that after the dose h only a fraction f < n of hits had occurred, i.e., the probability of survival, is given by:

$$N_h/N_o = \exp(-ah) \sum_{0}^{n-1} (ah)^f/f! \tag{80}$$

The second multiple-hit hypothesis of the target theory follows directly from the single-hit theory. If a population of organisms contains n sensitive targets that are inactivated exponentially, the viability of the organism is ensured if fewer than n targets are hit. The probability that all the targets in such a group of n becomes inactivated is given by:

$$P(n) = (1 - \exp(-kd))^n \tag{81}$$

since exp (−kd) is the probability that targets are not hit. This assumes that the rate, k, of occurrence of hits is the same for all the targets. Therefore, the assumption is not simply that n hits per organism are required, but that each of n particles or targets within the cell must be hit at least once.[61]

Both Equations 80 and 81 describe convex curves, characterized by an initial shoulder followed by a nearly exponential behavior at higher doses. The fitting procedures of experimental dose-effect curves by one or other of either equations are not critical. Discrimination between the two models would require a level of experimental accuracy that is not attainable in practice. It has been suggested that the n value should be referred to as the "extrapolation number", whether the multi-hit or the multi-target theory is considered.[62]

The multi-target theory suggests a likely explanation for the observed variations in recovery rate in different environmental conditions. In fact, it envisages microbial death as an end point of a process of gradually increasing damage as the applied dose increases, since the cell dies only when all the n vital sites are hit. Under suitable environmental conditions the cell can recover.

B. Heat Inactivation

As indicated above, convex survival curves are seldom reported in heat inactivation studies. Moats[57] succeeded in fitting convex survival curves based on a multi-target model he developed for heat-treated bacteria. However, according to his model the survival curve is expected to be characterized, as in the multi-target theory developed by Atwood and Norman,[61] by an initial lag (shoulder) followed by essentially exponential behavior over a wide range of intensities. The fundamental equation developed by Moats was

$$P(S) = \sum_{X=0}^{X_L-1} (N/X) \exp(-kt(N - X)) \cdot (1 - \exp(-kt))^X \tag{82}$$

where P(S) is the survival probability, X is the number of critical targets inactivated at any time t, X_L is the number of critical sites that must be inactivated to cause death, N is the total number of targets, and k is the rate constant for inactivation of individual targets. The values of k, N, and X_L can be obtained from experimental data. Solving simultaneously for k,

$$d/s = N(\exp(-kt) - 2\exp(-kt + kt_{50}) + \exp(-2kt_{50}))/(\exp(-kt) - \exp(-2kt)) \quad (83)$$

where d is deviation from the mean (i.e., $X - X_L$), $s = (Npq)^{1/2}$ where $p = X/N$ and $q = (N - X)/N$, and t_{50} is the time at which 50% of the population is killed.[54]

According to the model, cells of *Salmonella anatum* heated at 55°C can survive if plated on trypticase soy agar (a relatively rich medium) having about 7.8 critical sites inactivated, while if plated on basal medium (a less rich medium), only about 3.3 sites need to be inactivated to cause death.

Estimation of N and the procedure for calculating the rate constants is statistically difficult.[63] In the example quoted above, N ranges between 38 and 175 (see Reference 57). As pointed out by Moats himself, the model is unable to explain all survival curves and especially those with a tail.[56,57]

Alderton and Snell[64] developed an empirical expression which gave a reasonable fit to heat-treated bacterial spore survivor curves, showing a shoulder followed by an exponential decay:

$$\mathrm{Log}_{10}(N_o/N_t)^a = k't + C \quad (84)$$

where a is a constant characterizing the degree of resistance of the microorganism and C is a constant whose value increases with both treatment temperature and the sensitivity of the microorganism. Equation 84 allows the linearization of survivor curves obtained following treatment with heat, radiation, and disinfectants.[63,65] Alderton and Snell did not supply an explanation of the parameters a and C, nor of the inactivation mechanism.

C. Limitations of the Theory

The target theory was developed to explain the shoulder in survivor curves of irradiated microorganisms. The single-hit survivor curve is a special case of the theory. Nevertheless, the target theory does not explain other types of survival curves occurring in many radiation inactivation experiments, such as true sigmoid curves, continuously decreasing death rate curves, and curves with long tails. The merit of the theory is that it suggests a multiplicity of events leading to death as a possible general mechanism of microbial inactivation.

IV. CONTINUOUSLY DECREASING DEATH RATE CURVES

A. Experimental Evidence

Moats et al.[56] stated that " . . . examples of non-exponential survivor curves found in the literature are too numerous to list". Actually, as shown in Table 3, the list of only more prominent sigmoid or concave survivor curves reported in the literature is long. Nevertheless, some authors[8,50,66] seem to ignore factual evidence, proposing hypothetical biological or experimental reasons for deviation from exponential behavior, although their proposals are not convincing and they provide little in the way of experimental evidence. The early experimental evidence of Bigelow and Esty[5] showing many data incompatible with the exponential tenet and by Esty and Meyer's[6] data showing a *Clostridium botulinum* spores heat destruction curve with a tail lasting about 40 min seem to be disregarded.

The experimental evidence of deviations from exponential kinetics by many lethal agents is too widespread to be ignored.

Table 3
SOME REPRESENTATIVE NONLOGARITHMIC SURVIVOR CURVES OBTAINED WITH DIFFERENT MICROORGANISMS AND LETHAL AGENTS

Lethal agents	Microorganisms	Ref.
Low R.H.[a]	Yeasts	67
	Veg. bact.	68
Low R.H. + UV	Veg. bact.	69
Low R.H. + ethylene oxide	Veg. bact.	70
	Bact. spores	70
Heat	Veg. bact.	14—56,71—75
	Bact. spores	5—7,51,76—82
	Yeasts	83—86
	Molds	65,87
	Viruses, phages	88—92
	Toxins	93—95
	Enzymes	96,97
Heat + low R.H.	Veg. bact.	72
Heat + formaldehyde	Bact. spores	98
Heat + phenol	Bact. spores	99
Heat + hydrogen peroxide	Bact. spores	42
Heat + antimicrobials	Yeasts	100
Heat ± UV ± H_2O_2	Bact. spores	101
Heat + ultrasound	Bact. spores	102
Low temperature	Viruses	103
	Veg. bact.	104—106
Formalin, propiolactone	Viruses	88
Oxygen, ozone	Veg. bact., viruses	107,108
Phenol	Bact. spores	99
Iodine	Viruses	109
Alkalis, chloramines	Bact. spores	40
Chlorine dioxide	Veg. bact.	46
Benzoyl peroxide	Veg. bact.	110
Chlorine	Viruses	111
	Veg. bact.	111,112
Peracetic acid	Bact. spores	45
	Molds	113
Glutaraldehyde	Bact. spores	114
Visible light	Bact. spores	115
UV radiation	Veg. bact.	25,69,109,116, 117
	Yeasts	118
	Molds	25,118,119
	Bact. spores	25,101,120
Ionizing radiations	Bact. spores	31,32,121—129
	Veg. bact.	31,32,36,60, 116,117,130, 131
	Viruses	90,132

[a] R.H. = relative humidity.

B. Theoretical Aspects

A concave or biphasic survival curve suggests a phenomenon brought about by population heterogeneity. Chick's[4] proposal that heterogeneity in heat resistance could be responsible for a concave survivor curve seemed to be the only explanation for more than 70 years, although based on very little experimental evidence.

Microbial heterogeneity can explain biphasic survivor curves.[7,133-136] A mixture of two populations of organisms having different resistances to a lethal agent yields a biphasic survival curve. For example, mixing 10^7 spores of *C. botulinum* with 10^3 spores of PA 3679 and then subjecting the suspension to a temperature of 110°C, results in two survivor curves which intersect after about 15 min of treatment. Biphasic survivor curves could also occur when cells of a single species are composed of two populations with respect to their resistance to the lethal agent employed. Based on the same reasoning, multiphasic survivor curves, or continuously decreasing death rate curves (CDDRC), would be expected when heterogeneous populations are treated.[137] The first problem raised by several authors was whether the distribution of resistance among individuals in a population was permanent ("innate heterogeneity" theory)[7,8,138-140] or if it was acquired during the treatment ("adaptation model").[141-143] The second problem was which type of probability distribution could explain different shapes of the survivor curves.

C. Mathematical Models

Han et al.[135] developed a model for both the innate heterogeneity hypothesis and the adaptation hypothesis. The following equation was derived for the former:

$$\mathrm{Log}_{10}(N_t/N_o) = -Kt + (s^2/2)\,t^2 \tag{85}$$

where K is the most probable value of the destruction rate, s is the standard deviation, and t is the treatment time. The heat adaptation approach leads to the following equation:

$$\mathrm{Log}_{10}(N_t/N_o) = -K_oX((1 - a)\,t - ab(\exp(-t/b) - 1)) \tag{86}$$

where K_o is the initial rate of destruction, a is a constant representing the maximum amount of resistance attainable for a unit amount of destructive power, b is a constant representing the rate of development of resistance, and t is the time. By applying the two models to some bacterial inactivation curves, the authors concluded that curvilinearity in the survival curves resulted from the development of resistance during the treatment, rather than from innate heterogeneity.

Sharpe and Bektash[136] modified the models developed by Han et al.[135] by utilizing other types of probability distributions of resistance in the population, including the normal, γ, shifted γ, and a modified Poisson distribution, and suggested that a combination of the innate heterogeneity and adaptation models might be appropriate. They proposed that the distribution of the initial rate of destruction (K_o) could be represented by any distribution having a probability $P(K_o < 0) = 0$, that the life of a cell follows an exponential distribution with mean 1/K, and that the state of the cells (i.e., living or dead) is independent. These considerations lead to a probability of survival S(t) at time t of the following form:

$$S(t) = \mathrm{Log}(N_t/N_o) = \mathrm{Log}\ L_f(t) \tag{87}$$

where $L_f(t)$ is the Laplace transform of the density function of K at time zero, that is for a normal distribution,

$$L_f(t) = \exp(-K_ot + (s^2t^2)/2) \tag{88}$$

for a γ distribution,

$$L_f(t) = (1 + t/\lambda)^{-r} \tag{89}$$

for a shifted γ distribution,

$$L_f(t) = \exp(-at)(1 + t/\lambda)^{-r} \qquad (90)$$

where a is the minimum value of the rate of destruction and for a modified Poisson distribution,

$$L_f(t) = \exp(-K_o(1 - a)\, t + K_o ab(\exp(t/b) - 1) \qquad (91)$$

Sharpe and Bektash[136] concluded that it is not possible to distinguish between the two possibilities (innate heterogeneity or development of resistance) on the basis of the survival data alone; the heat adaptation model of Han et al.,[135] for instance, has been shown to be equivalent to an innate heterogeneity model with a modified Poisson distribution. Nevertheless, from the mathematical analysis carried out by the authors, it can be argued that all concave survivor curves can be reasonably explained by the innate heterogeneity theory.

Two objections can be made to the purely phenomenological models described above: first, the adaptation model cannot be established if the acquisition of resistance during the treatment is not permanent; second, the distribution of resistance in an untreated population can be demonstrated experimentally. Furthermore, none of the models examined is able to explain the tailing phenomenon. As shown later, the experimental evidence is contrary to both models.

Brannen[144] developed a model based on the assumption that survival depends on a number of subsystems, whose functionality is affected by heat. The model yields the four classical types of survivor curves, although ''testing of the model is extremely difficult''[144] and the relationship between heat resistance and water content of the environment is not explained.

D. Evidence Against Suggested Models

As pointed out by several authors, the occurrence of small numbers of very resistant individuals could be regarded as a normal feature of a population of microorganisms. To verify the assumption that CDDR curves and tailing result from the type of the distribution of resistance in the population, at least two types of experiments must be done: (1) particles surviving more drastic treatments must be assayed in order to show if their resistance is greater than that of the majority of individuals in the population and (2) the resistance of decreasingly smaller fractions of the population must be assayed to determine whether the CDDR curves become progressively exponential as cell counts decrease.

The first type of experiment has been performed by few authors, and all failed to show survivors of greater resistance (to heat, radiation, or disinfectant) than in the parent population.[14,56,73,138,145] The second type of experiment has been performed by more authors. Bigelow and Esty[5] reported heat resistance data obtained using thermophilic bacterial spores treated at temperatures ranging from 100 to 140°C. They used mother suspensions containing more than 10^5 spores per sample and diluted these down to 3 spores per sample. Surprisingly, since it was unexpected both on the basis of a hypothetical distribution of the resistance among individuals in the spore population and on the basis of expected exponential inactivation, it was found that the time required to destroy 90% of the spores increased as particle concentrations (N_o) decreased. More than 80% of the assays carried out by Bigelow and Esty[5] showed such an effect. The decreasing death rate found as spore concentrations decreased was increasingly evident (and obviously statistically more reliable) as treatment temperatures were lowered. This phenomenon escaped the attention of many researchers. Many authors followed the suggestion of Stumbo et al.[146] and averaged the D_T values they obtained, disregarding the actual meaning of the decrease in death rate with increasing treatment time. Reed et al.[147] found a similar phenomenon in heat destruction rate studies on PA 3679 spores. Pflug and Esselen[148] found increased resistance of PA 3679 spores at all 14 temperatures tested (ranging from 112.8 to 148.9°C), as treatment time increased. Kempe et al.[149] found a linear relationship between the logarithm of *Clostridium botulinum* 62A spores and a dose of γ-radiation. At spore concentrations ranging from $4 \cdot 10^4$ to $4 \cdot 10^2$ the decimal reduction dose ranged from 0.6 to 0.8 Mrad; the D_{90} increased to 1.5 Mrad at lower concentrations and reached as high as 12.0 Mrad at a spore concentration of four per ten samples.

Amaha[12] found a linear relationship between Log_{10} time and $Log_{10}N_o$ using spores of *C. sporogenes, Bacillus megaterium,* and *B. natto* treated at temperatures ranging from 105 to 120°C. Using PA 3679 spores Casolari[82] found an increase in D_T value as spore concentrations decreased from $9 \cdot 10^6$ to $1.2 \cdot 10^0$, employing ten different media for the recovery of treated spores. Using five *C. botulinum* strains (types A, B, and E) and six PA 3679 strains, the extent of inhibition of growth of the vegetative cells brought about by nitrite-dependent-compounds (inhibitory substances present in heat treated solutions containing nitrite) was found to be linearly correlated with the logarithm of the initial cell concentration, ranging from 4 to 10^6 per sample.[150] Analogous results, although unrecognized or disregarded, were obtained by Greenberg,[151] Crowther et al.,[152] and Roberts and Ingram[153] among others, using heat treated substrates containing nitrite. Greater heat resistance by low concentrations of yeasts was shown by Williams (see Morris[85]) and Casolari and Castelvetri;[86] Campanini et al.[75] found the same phenomenon using *S. faecalis.*

Spores of *B. polymyxa* ($4 \leq N_o \leq 10^4$ per experimental unit) and vegetative cells of *Staphylococcus aureus* ($4 \leq N_o \leq 200/m\ell$) and *E. coli* ($0.1 \leq N_o \leq 10^2$ cells per 10 mℓ) were treated with γ-radiation from a ^{60}Co source using three dose rates (from 1 to 22 krad/min).[154] The results showed clearly that the initial concentration of particles affected survival probability at all the dose rates tested.[155] When *S. aureus* was irradiated in solutions containing cysteine at 10^{-3} *M*, the D_{90} doubled at $N_o = 4.6$ cells per 10 mℓ (D_{90} (200 cells) = 31 krad, D_{90} (4.6 cells) = 73 krad) and was three times higher at a cell concentration of 0.93/10 mℓ (D_{90} = 214 krad).[154,155]

According to the above observations, CDDR curves neither result from heterogeneity in the resistance of individuals in a population, nor from the acquisition of resistance during treatment. Therefore, a different hypothesis must be formulated.

V. APPROACH TO A GENERAL MODEL

Some years ago a model was devised[155] which related in some way concepts suggested by Ball and Olson[7] about the likely mechanism of microbial heat inactivation. The model was based originally on what might be envisioned to occur in the process of heat inactivation, although, as will be shown later, it applies to radiation and chemical inactivation processes as well.

A. Heat Inactivation

The basic reasoning was that as shown experimentally, a single factor is of paramount importance in the heat inactivation process. This is the water content of the environment. A suspension of microbial cells in aqueous medium can be regarded as a biphasic system consisting of about $3 \cdot 10^{22}$ water molecules and less than 10^9 microbial particles per milliliter. Energy supplied to a system will be taken up by the more concentrated components of the system and then transferred by collision to less concentrated ones. Accordingly, kinetic energy supplied to a microbial suspension is expected to be taken up by water molecules and then transferred to microbial cells. Brownian movement of particles results from this collision process. If the energy transferred between the particles is sufficiently high, the physicochemical structure of the microbial particle is damaged. If the damage is great enough the particles lose their ability to functionally relate with their environment and become unable to multiply, viz., they die. A more detailed hypothesis about the death mechanisms has been reported elsewhere.[155]

Let the survival probability $P(S) = C_t/C_o$, where C_o and C_t represent the concentration of living organisms initially and after a treatment time t, respectively. Based on the experimental evidence outlined above, the probability with which particles elude collisions (q) with water molecules carrying lethal energy E_d can be regarded as inversely related to the living particle concentration in the suspension at the time t:

$$q = 1/C_t \tag{92}$$

The probability $P_o(T)$ with which particles elude lethal collisions at temperature T is expected to depend on the frequency Pc of collision at the temperature T, so that:

$$P_o(T) = q^{Pc} \tag{93}$$

and during t min at the temperature T:

$$P_o(T) = q^{tPc} \tag{94}$$

The collision frequency Pc depends on both the probability of there being a given number of water molecules with more than E_d energy, that is $P(n_E)$, and on the probability P(h) that available n_E molecules strike microbial particles, so that:

$$Pc = P(n_E)\,P(h) \tag{95}$$

Taking into account the relative size of microbial particles (about 10^{11} times greater than a water molecule), the probability P(h) almost equals the probability of having n_E molecules per unit volume, so that Equation 95 can be rewritten:

$$Pc = (P(n_E))^2 \tag{96}$$

The $P(n_E)$ value comes from the Maxwellian distribution of energy from which, in the simplified form (Equation 60), the number of molecules carrying more than E_d energy present in 1 mℓ of water will be

$$P(n_E) = (6.02295 \cdot 10^{23}/18)\exp(-E_d/RT) \tag{97}$$

so that:

$$Pc = M = (6.02295 \cdot 10^{23}/18)^2 \exp(-2\,E_d/RT) \tag{98}$$

that is

$$M = \exp(103.7293 - 2E_d/RT) \tag{99}$$

It follows that the survival probability after time t at temperature T is described by the equation:

$$P(S) = C_t/C_o = q^{Mt} = (1/C_t)^{Mt} = C_t^{-Mt} \tag{100}$$

Dividing by C_t:

$$C_o = C_t^{(1+Mt)} \tag{101}$$

or

$$C_t = C_o^{(1+Mt)^{-1}} \tag{102}$$

that is:

$$C_t = C_o^{(1+t)\ \exp(103.7293 - 2\ E_d/RT)^{-1}} \tag{103}$$

To fit experimental data using Equation 103 we must know values for C_o, C_t, and t. With these values for a single temperature, the value of M can be obtained from Equation 102:

$$M = ((\text{Log } C_o/\text{Log } C_t) - 1)/t \tag{104}$$

and hence the E_d value:

$$E_d = 0.5\ RT(103.7293 - \text{Log } M) \tag{105}$$

The E_d value is the most important single parameter characterizing the heat resistance of microbial cells in defined environmental conditions. It can be expected that for a given microorganism, E_d will depend on the environmental conditions pertaining after heat treatment.

Knowing the E_d for a given microorganism, the inactivation curves at any temperature and at any environmental water content can be obtained by simple computation.

1. Expected Shape of Survivor Curves

Survivor curves obtained by plotting $\text{Log}_{10}C_t$ against time t (min) are expected to be, according to the model, fundamentally concave. Nevertheless:

1. At a given temperature T, survivor curves of microorganisms having high E_d are expected to be nearly exponential (i.e., statistically indistinguishable from an exponential decay curve); those of organisms having intermediate values of E_d are expected to be concave (i.e., of the CDDR type); and survivor curves of microorganisms having low E_d are expected to be nearly exponential initially, followed by a phase of CDDR type and finally tailing (Figure 5).
2. A population of organisms having a given E_d is expected to yield nearly exponential inactivation curves at low temperature, concave (CDDR type) survivor curves at intermediate temperatures, and curves nearly exponential at first (short treatment times), followed by a CDDR phase and then tailing (Figure 6).
3. Depending on the concentration of cells, survival curves at a given temperature T are more concave (high concentration) or less concave (low particle concentration) (Figure 7). Equation 103 agrees quite well with experimental data, as already shown.[155]

Pflug and Esselen[148] reported a total of 95 experimentally determined decimal reduction times obtained by treating PA 3679 spores at 14 temperatures; the relationship obtained by plotting $\text{Log}_{10}D_T$ against temperature yields $\text{Log}_{10}D_T = 12.986 - 0.107\,T$ ($r = -0.9992$). Figure 8 shows the ratio between experimental D_T values and the D_T expected by interpolation, using the above equation, together with those expected from the model. The computation of D_T values as performed by authors was possible only if a fraction of heat treated samples was sterile, since in order to obtain the number of survivors, they used the first term of the Poisson distribution, known as the Halvorson and Ziegler formula:

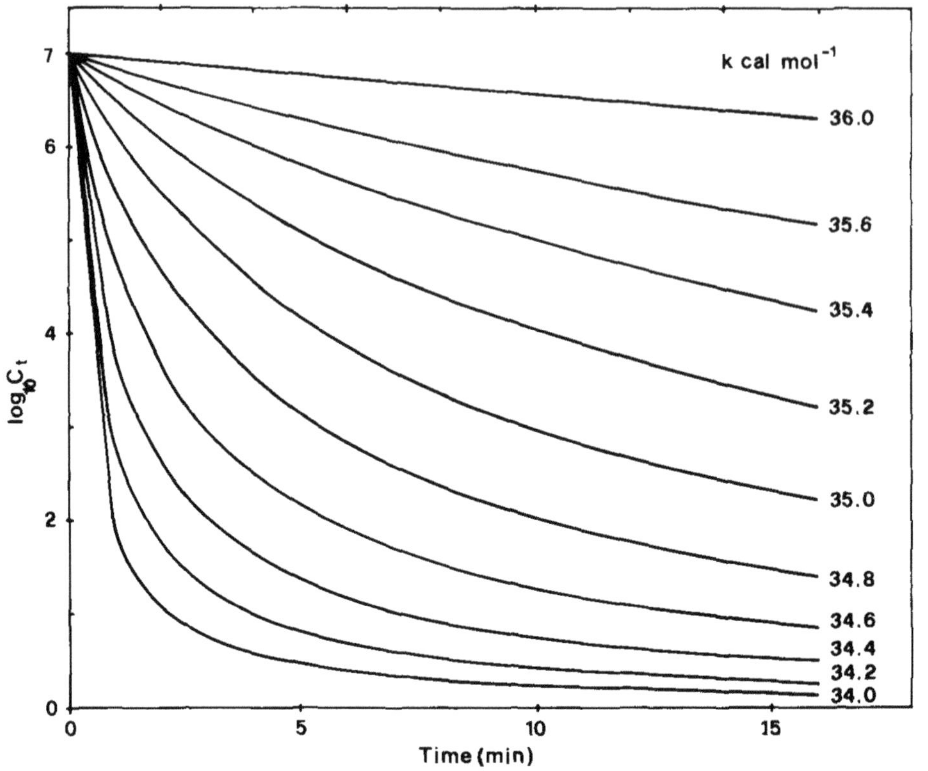

FIGURE 5. Predicted shape of survivor curves of microorganisms heated at the temperature T, as a function of the lethal energy E_d (kcal/mol), according to the general model. Correlation coefficients of linear regression obtained from ten pairs of data were r = −0.9994, −0.9902, or −0.9485, for E_d = 36, 35.4, or 35 kcal/mol, respectively. C_t is the concentration of organisms surviving t (min).

$$\overline{N}_t = \text{Log}(A/H) \tag{106}$$

where $\overline{N}_t$ is the average number of survivors per sample, A is the total number of treated samples, and H is the fraction of sterile samples. To use the model, the first term of the Poisson distribution was used in order to obtain a C_t value and to compute an M value from Equation 104. A parameter analogous to the decimal reduction time, subsequently called P(10,T), was obtained from Equation 100:

$$P(10,T) = 1/M \; \text{Log}_{10}C_t \tag{107}$$

To use the above equation, a C_t value corresponding to 37% sterility of treated samples was chosen for all temperatures; Pflug and Esselen[148] used N_t values derived from sterility fractions ranging from 4 to 96%. As can be seen from Figure 8, the experimental data are quite scattered around the interpolated values. Nevertheless, both experimental D_T values and those predicted by the model show a defined trend; that is, they are not equally distributed about values expected from the regression obtained using all data, although they show a defined concavity around interpolated values obtained by assuming that z is constant. Such behavior is pertinent to the controversy regarding the linear dependence of inactivation rate against temperature.[156,157] The question arises as to whether the Arrhenius activation energy or the z value can be regarded as being constant. The moderate concavity obtained by plotting microbial heat destruction rates against T^{-1} ($\text{Log}_{10}D_T$ vs. T^{-1}) is statistically indistinguishable from the plot of $\text{Log}_{10}D_T$ against T, taking into account the low level of experimental accuracy

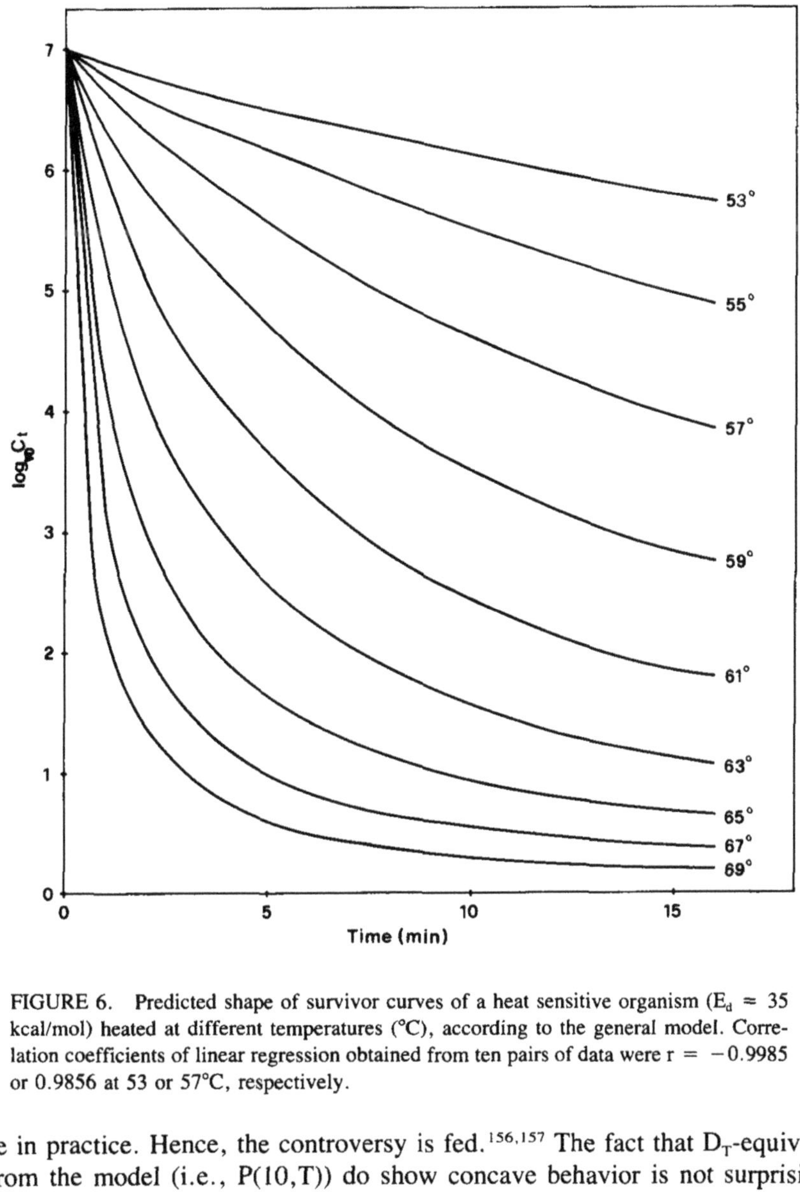

FIGURE 6. Predicted shape of survivor curves of a heat sensitive organism (E_d = 35 kcal/mol) heated at different temperatures (°C), according to the general model. Correlation coefficients of linear regression obtained from ten pairs of data were r = −0.9985 or 0.9856 at 53 or 57°C, respectively.

attainable in practice. Hence, the controversy is fed.[156,157] The fact that D_T-equivalent data arising from the model (i.e., P(10,T)) do show concave behavior is not surprising, since the model requires that E_d must be constant; the unexpected evidence coming from the experimental data reported by Pflug and Esselen,[148] on the contrary, suggests that Log D_T is linearly correlated with T^{-1} and not with T (i.e., it is lethal energy which is constant, not z). It follows that a plot of $\text{Log}_{10}D_T$ against T is not appropriate.

2. *Tailing-Off*

According to Equation 107, the inactivation rate decreases as C_t decreases. At a low C_t level following the M level in the environment (i.e., temperature, free water, etc.), the inactivation rate is expected to be very low and it approaches zero as C_t approaches unity. Tailing can be regarded as a phenomenon produced by the increasingly low probability of collision between water molecules having more than E_d energy and microbial particles. If microbial particles are about 10 μm apart on the average (as in suspensions containing 10^9 particles per milliliter), water molecules carrying lethal energy can be expected to have a greater probability of striking the particles than when the particles are more than 1000 μm

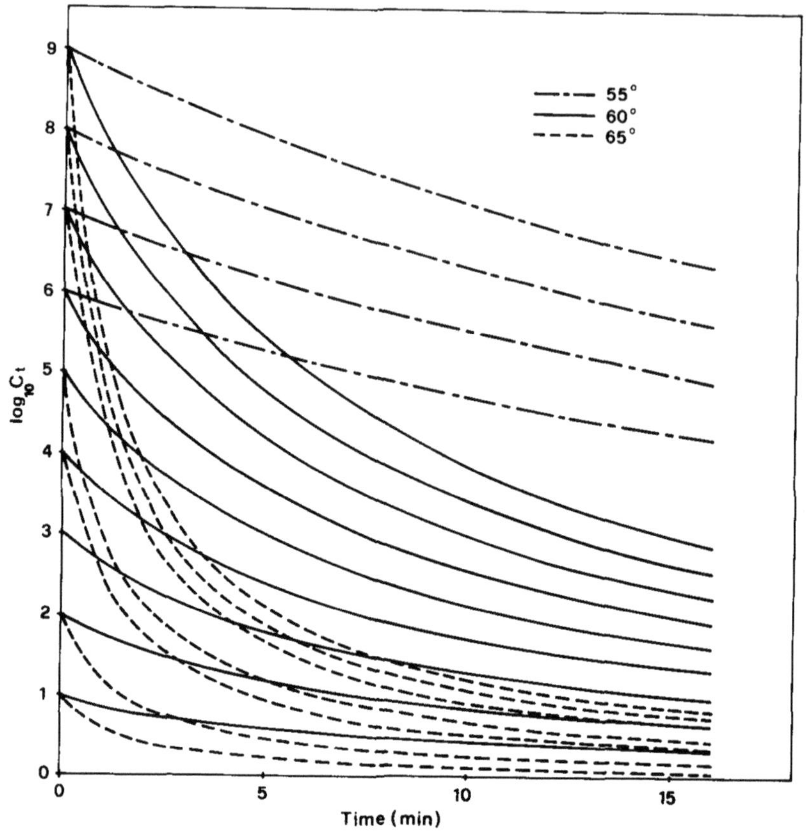

FIGURE 7. Predicted shape of survivor curves of a heat sensitive microorganism (E_d = 35 kcal/mol) as a function of the concentration of untreated organisms (C_o) and of the temperature, according to the general model. The correlation coefficients of linear regression from ten pairs of data obtained at 55°C are all greater than −0.99.

apart (as in suspensions containing less than 10^3 particles per milliliter). Actually, a water molecule having high energy behaves like a long-lived bullet, able to overcome a large number of collisions with other water and/or solute molecules, while traveling through the medium, keeping enough energy to kill living particles it meets along its path. Nevertheless, in environmental conditions which result in more tailing-off, the concentration of M molecules is very low. In heat inactivation curves, the tail appears after shorter treatments as temperature increases. The probability of finding the tailing-off of survivor curves decreases as temperature increases, since microbial particles are inevitably struck with increasing frequency and also by molecules having less than E_d energy while having a value high enough to damage the particles. The probability of occurrence of molecules having enough energy to damage the particles is expected to increase with temperature, according to the Maxwell-Boltzman distribution of energy. It follows that a fraction of particles could become incapable of growing, although surrounded by suitable environmental conditions, being extensively damaged not by molecules carrying more than E_d energy, but by those carrying less than E_d energy. Such an event is not accounted for by the model developed, while it can be expected to occur following the premise of random collisions. However, these considerations concern almost exclusively the last living particle, since the model predicts that inactivation curves tend to become exponential as temperature increases owing to the ensuing increase of M value. According to the model, the death of the last particle cannot be expected to occur.

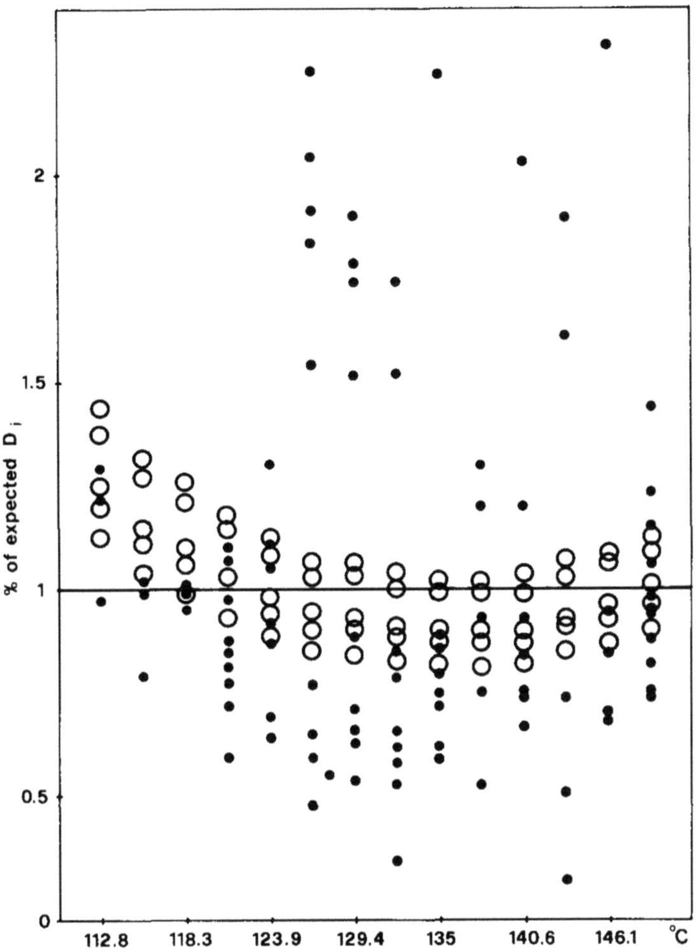

FIGURE 8. Relationship between temperature and (1) the ratio of single experimental D_T values ($D_T(e)$) reported by Pflug and Esselen[148] to those obtainable from the regression equation ($D_T(i)$) calculated using the 95 experimental data (r = −0.9999), (2) the ratio of P(10,T) (the parameter analogous to D_T, from the general model) to $D_T(i)$. Full circles = $D_T(e)/D_T(i)$; empty circles = $P(10,T)/D_T(i)$. In using $P(10,T) = 1/M \, Log_{10}C_t$, five experimental survival fractions were chosen at random and M values calculated using Equation 104.

Tailing occurs at decreasingly low particle concentrations, as temperature increases. There are intermediate temperatures at which tailing occurs with a greater probability, as well as extreme ones (low or high temperatures) at which inactivation curves resemble a straight line.

Very low survivor values are often disregarded on the basis that they are not statistically reliable.[158] On the other hand, they are accepted for exponential inactivation curves at values as low as 10^{-12}.

3. Inactivation Rate, Temperature, and Energy

As noted previously, a relevant parameter employed to define heat resistance in microorganisms is z (Equation 15), related to the temperature coefficient, Q_{10}, by the relationship expressed in Equation 16.

According to Equations 99 and 107, the inactivation rate increases as the M value increases, and M increases with temperature. The model suggested is able to supply an exact definition of z and Q_{10} in terms of M: the inactivation rate changes Q_{10} times for each 10°C, since the M value changes Q_{10} times for each 10°C; the inactivation rate changes ten times for each change of z degrees, since the M value changes ten times for each change of z degrees.

According to the model (Equation 107):

$$(P(10,T))^{-1} = M \, Log_{10}C_t \tag{108}$$

then:

$$Q_{10} = M_{(T+10)} \, Log_{10}C_t/M_T \, Log_{10}C_t$$
$$= M_{(T+10)}/M_T \tag{109}$$

It follows that:

$$Q_{10} = \exp((2 \, E_d/R)(10/(T + 10)T)) \tag{110}$$

and from Equation 99,

$$Log \, Q_{10} = (2 \, E_d/R)(10/(T + 10)T) \tag{111}$$

Therefore, E_d can be obtained as a function of Q_{10}:

$$E_d = (Log \, Q_{10}/10)(R/2)((T + 10)T) \tag{112}$$

and from Equation 16,

$$Q_{10} = \exp(23.03/z) \tag{113}$$

so that,

$$E_d = (Log \, 10) \, R \, T(T + 10)/2 \, z \tag{114}$$

and

$$E_d = R \, T(T + 10) \, Log \, Q_{10}/20 \tag{115}$$

since, according to Equation 113,

$$z = (10 \, Log \, 10)/Log \, Q_{10} \tag{116}$$

Accordingly, given $Q_{10} = z = 10$ in the temperature range 110 to 120°C and following Equations 112 and 114, E_d will equal 34459.6348 cal/mol. Therefore, using Equation 99, the value of M will be $5.4208 \cdot 10^5$ at 110°C, a value which is $z = Q_{10} = 10$ times lower than the value of M at 120°C where it is $5.4208 \cdot 10^6$. Similarly, given a Q_{10} (a) = 22 (i.e., z = 7.4449) at $60° \leq T \leq 70°C$, and a Q_{10} (b) = 16 (i.e., z = 8.305) at $110° \leq T \leq 120°C$, from Equations 112 and 114, E_d (a) = 35107.2379 and E_d (b) = 41493.5348. In the former case $M_{70°C} = 2.1196$, which is Q_{10} (a) = 22 times greater than $M_{60°C}$ which is equal to 0.096. In the latter case, $M_{120°C}$ is 0.082 which is Q_{10} (b) = 16 times greater than $M_{110°C}$ which has a value of $5.1178 \cdot 10^{-3}$. At the same time, at 60°C + z = 67.449°C,

Table 4
EXPECTED VALUES OF z AND OF Q_{10} AS A FUNCTION OF THE TEMPERATURE AND OF THE LETHAL ENERGY E_d, FOLLOWING THE GENERAL MODEL[155]

E_d	z	Q_{10}	z	Q_{10}	z	Q_{10}	z	Q_{10}
	Heat-Sensitive Organisms							
	Temperature range (°C)							
	55—65		**65—75**		**75—85**		**85—95**	
33	7.69	19.95	8.16	16.80	8.64	14.35	9.14	12.42
34	7.47	21.85	7.92	18.30	8.39	15.56	8.87	13.40
35	7.25	23.92	6.70	19.93	8.15	16.87	8.62	14.47
36	7.05	26.19	7.48	21.71	7.92	18.28	8.38	15.61
37	6.86	28.68	7.28	23.65	7.71	19.82	8.15	16.85
	Heat-Resistant Bacterial Spores							
	Temperature range (°C)							
	95—105		**105—115**		**115—125**		**125—135**	
38	8.38	15.60	8.84	13.54	9.30	11.88	9.78	10.52
39	8.17	16.77	8.61	14.50	9.07	12.68	9.53	11.20
40	7.96	18.03	8.39	15.53	8.84	13.53	9.29	11.91
41	7.77	19.38	8.19	16.64	8.62	14.45	9.07	12.67
42	7.58	20.84	7.99	17.82	8.42	15.42	8.85	13.48

Note: E_d = kcal/mol.

M equals 0.963 which is 10 times higher than $M_{60°C}$ and at 120°C − z = 111.695°C, M equals 0.0082 which is a value 10 times lower than that of $M_{120°C}$.

Equation 110 shows the rate of change of Q_{10} as a function of E_d and temperature. The rate of change of z as a function of E_d and temperature is

$$z = (\text{Log } 10)(T + 10)\ T\ R/(2\ E_d) \qquad (117)$$

Taking into account the inactivation rate of heat-sensitive microbial particles destroyed in measurable times at 60°C it can be computed that according to the model, the expected Q_{10} values must range between about 18 and 26 at 60°C ≤ T ≤ 70°C, so that 8°C ≤ z ≤ 7°C is expected in the same temperature range, as shown in Table 4. For more resistant bacterial spores, destroyed in measurable times at 110 to 120°C, the expected values of Q_{10} range between about 13 and 20, so that 7.5°C ≤ z ≤ 9°C is expected in the same range of temperature. As can be seen in Table 4, the expected value of z increases as temperature increases; in the same temperature range, it decreases as E_d value increases (i.e., increasing the resistance of the organism). The opposite is true with Q_{10}.

It follows that the energy, E_d, required to inactivate less resistant organisms ranges between about 33 and 37 kcal/mol, while E_d values required to inactivate more resistant organisms range between about 38 and 44 kcal/mol. Thus, less energy is required to kill less resistant organisms, and vice versa, a result quite reasonable but in opposition to what is predicted by the classical thermodynamic treatment of exponential kinetics.

The number of water molecules having E_d values predicted by the model is quite reasonable, in opposition to that predicted from thermodynamic treatments. At an intermediate E_d value of 35 kcal/mol, for instance, for less resistant organisms, the M values range between about 0.1 and 3/mℓ at 60 and 70°C, respectively; given E_d = 40 kcal/mol for more resistant organisms, the M value equals 0.25 at 110°C and 3.8/mℓ at 120°C.

4. Inactivation Rate and Water Content of the Environment

The single most relevant environmental condition affecting heat resistance of microbial particles is the amount of water contained in the medium in which microorganisms are treated.[159-162] This phenomenon can be explained simply by the model proposed. The M value is linked to the number of water molecules per milliliter or gram of substrate, according to Equation 98, which can be rewritten as:

$$M = \exp(A - 2\,E_d/RT) \tag{118}$$

where:

$$A = 94.5190 + 2\,\mathrm{Log}\,W \tag{119}$$

and 94.519 is the natural logarithm of the squared number of water molecules per gram of medium containing 1 g of water and W = g of water in 100 g of medium. Using Equations 118 and 119 we may obtain the value of M at any given water content ($W \leq 100$). Obviously, as the water content decreases, so does the M value. From Equations 118 and 119,

$$\mathrm{Log}\,M_w = 94.519 + 2\,\mathrm{Log}\,W - 2\,E_d/RT \tag{120}$$

where M_w is the expected reduced M value following a decrease in the W value. A decrease in water content of the environment is expected to be equivalent to a decrease in temperature. The temperature T_e (equivalent temperature) corresponding to an M value lower than the one expected in a fully hydrated environment can be obtained by substituting M obtained from Equation 120 into Equation 99, i.e.,

$$T_e = 2\,E_d/R(103.7293 - \mathrm{Log}\,M_w) \tag{121}$$

Since T_e is lower than T (the true temperature in a fully hydrated environment), it follows that decreasing the water content in the medium results in an increase in microbial resistance, viz., a decrease in water content is equivalent to a decrease in temperature. As shown in Figure 9, a temperature of 120°C in a fully hydrated environment is equivalent to about 87.6°C in a medium containing 1 g of water in 100 g of medium.

Taking into account the z(M) and Q_{10} (M) expected values for both sensitive and resistant particles, a change of water content from 100 to 1 g in 100 g of medium at a given temperature T (°C + 273.15), the inactivation rate decreases about 1000 times (Figure 9).

The following equation:

$$T' = 2\,E_d/(R(94.519 + 2\,\mathrm{Log}\,W - \mathrm{Log}\,M_T)) \tag{122}$$

can be used to compute the temperature T′ that must be reached to obtain a P(10,T) equal to that expected at the temperature T in a fully hydrated environment. As shown in Figure 9, a 100 times change in W requires that T reaches about T′ = T + 40°C (letting E_d = 40 kcal/mol).

The agreement between resistance data at different water contents in the medium, and those predicted by the model, has already been shown.[155]

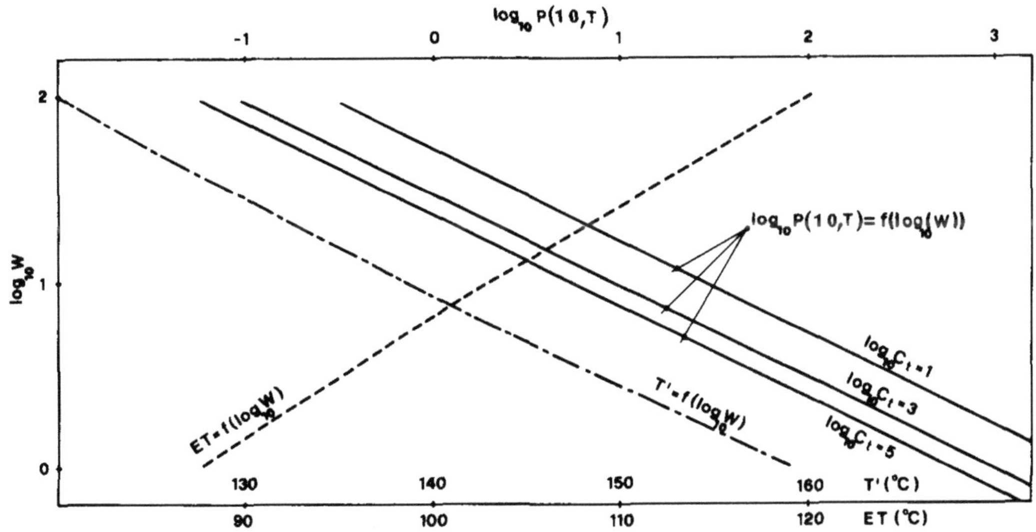

FIGURE 9. Relationship between M value in a fully hydrated environment at 120°C (for E_d = 40 kcal/mol) and (1) the temperature ET at which an equivalent value of M is expected to occur, according to the general model, as water content of the environment decreases (ET = f (Log_{10}W), where W = g of water/100 g of medium), (2) the temperature T′ that must be reached, according to the model, to obtain a value of M equal to that occurring in a fully hydrated environment (T′ = f (Log_{10}W)). Relationships of the type Log_{10} P(10,T) = f (Log_{10} W) represent the expected decrease of P(10,T) as water content of the environment also decreases, and were computed using three C_t values for illustrative purposes.

Several workers have pointed out that heat inactivation rates in dry environments have greater z values than those obtained in fully hydrated ones. According to Fox and Pflug,[78] it is reasonable to suppose that cells subjected to high temperatures in dry environments lose water. The increase of z in dry environments can be explained, according to the model, accepting the suggestion of these authors. Otherwise, as E_d is a constant for any given microorganism, z must change thereby changing the water content of the environment. Let Q'_{10} and z′ be values of Q_{10} and z expected when the water content at the temperature T is W_T and that at the temperature T + 10°C is $W_{T+10°C}$. Then:

$$\text{Log } Q'_{10} = 2(\text{Log } W_{T+10°C} - \text{Log } W_T) + (2E_d/R)(10/T(T + 10°C)) \quad (123)$$

so that:

$$z' = 10 \text{ Log } 10/(2(\text{Log } W_{T+10°C} - \text{Log } W_T) + (2E_d/R)(10/T(T + 10)) \quad (124)$$

Table 5 shows the fraction of water that must be lost in dry conditions by a 10°C increase in treatment temperature in order to obtain z values higher than those expected in moist conditions (W = 100 g water/100 g of medium). As can be seen, a loss of only about 50% of the water content, coming from a lower to a higher temperature, is required to yield very high z values (very low Q_{10} values), somewhat independently of the temperature range and/or the E_d value.

B. Radiation Inactivation

Survivor curves of microorganisms treated with radiation can be convex, sigmoid, or concave. A function like that suggested above for heat inactivation (Equation 102) can describe these types of survivor curves (Figure 10):

Table 5
LOSS OF WATER PREDICTED BY THE GENERAL MODEL, TO OBTAIN HIGH z (LOW Q_{10}) VALUES, AS A FUNCTION OF THE TEMPERATURE RANGE AND OF THE ENERGY E_d (kcal/mol), IN MICROBIAL INACTIVATION KINETICS

E_d = 35,000 (Heat-Sensitive Organisms)

		Temperature ranges (°C)			
z	Q_{10}	55—65	65—75	75—85	85—95
		Percent loss of water			
10	10.00	35.34	29.17	23.00	16.85
15	4.64	55.95	51.74	47.54	43.35
20	3.16	63.64	60.17	56.69	53.24
30	2.15	69.99	67.17	64.24	61.40
40	1.78	72.73	70.13	67.52	64.93
50	1.59	74.26	71.80	69.34	66.90

E_d = 40,000 (Heat-Resistant Bacterial Spores)

		Temperature ranges (°C)			
z	Q_{10}	120—130	130—140	140—150	150—160
		Percent loss of water			
10	10.00	11.20	5.57	0.02	[a]
15	4.64	39.50	35.67	31.88	28.16
20	3.16	50.06	46.90	43.78	40.71
30	2.15	58.78	56.17	53.59	51.06
40	1.78	62.55	60.18	57.84	55.54
50	1.59	64.65	62.41	60.20	58.02

[a] z value is necessarily higher (Q_{10} lower) than 10.

$$C_d = C_o^{(1+SD)^{-1}} \tag{125}$$

where C_d is the concentration of microorganisms surviving the dose d (Mrad) of radiation, D is the squared dose of radiation, and S is a sensitivity parameter, specific to the microorganisms employed. Experimental radiation survival curves can be fitted, according to Equation 125.[155]

Following the model, lethal effects of radiations are expected to be mainly "indirect". Radicals produced by radiation in aqueous media are expected to damage microbial structure, in particular surface structures, as occurs with several lethal agents.[53,55,163-167] Microbial resistance to radiation is higher in microorganisms with a higher content of –SH groups.[168] Let the maximum level of –SH/cell be about 10^8. If S in Equation 125 equals $10^8/G$, where G is the expected –SH content of a single microorganism, or more precisely the –SH fraction present at the surface of a microbial cell, experimental survival curves can be predicted quite accurately. Figure 11 shows the relationship between total –SH content/colony-forming-unit for different microorganisms and their decimal reduction doses[168] as compared with the analogous parameter P(10,R)′, i.e., the first decimal reduction dose, obtained from the model:

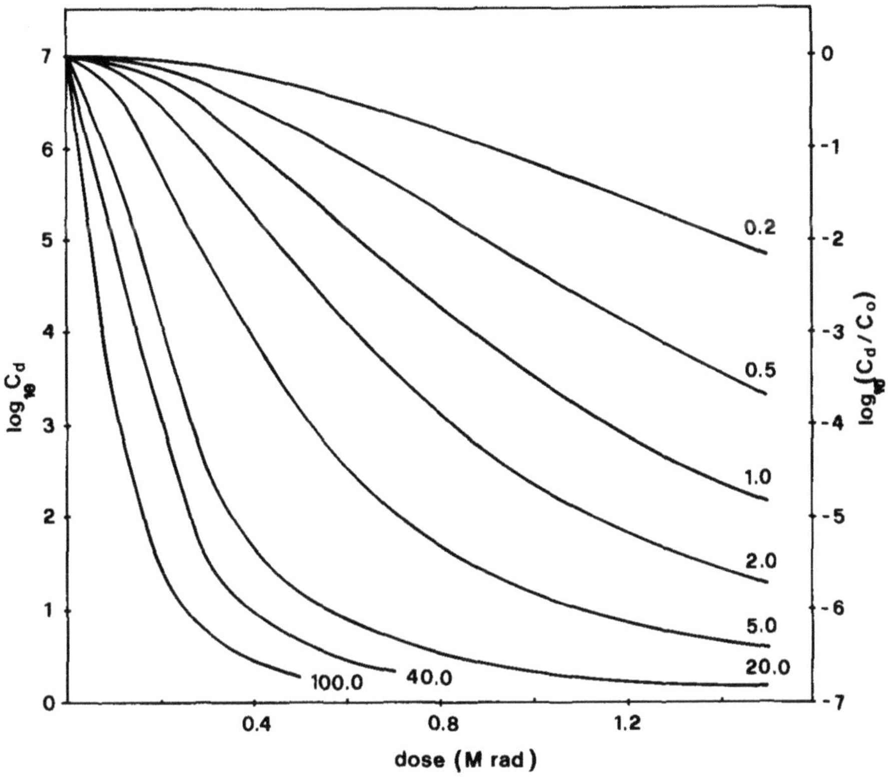

FIGURE 10. Shape of survivor curves of microorganisms treated with ionizing radiations, predicted by the general model, as a function of the value of the ratio $S = 10^8/(-SH)$, where (–SH) is the number of surface –SH groups/cell. Values of $0.2 \leq S \leq 100$ were selected since (–SH) content of most bacteria lies in this range.

$$P(10,R)' = ((\mathrm{Log}_{10}C_o/\mathrm{Log}_{10}(C_o/10)) - 1)/S \tag{126}$$

As can be seen, the agreement is quite satisfactory, taking into account the uncertainties associated with the evaluation of the –SH content per cell[168] and mostly the approximation made for computation purposes (i.e., 1/4 of the –SH content per cell was assumed to be at the surface of the cells, on the basis of data reported by Bruce et al.[168]). The high resistance of *Micrococcus radiodurans* could result from the exceptionally active repair system of this organism.[169]

C. Chemical Inactivation

Microbial inactivation kinetics by chemical compounds can be expected to be described by a function of the form:

$$C_t = C_o^{(1+Q\ t)^{-1}} \tag{127}$$

and

$$C_q = C_o^{(1+Q\ S)^{-1}} \tag{128}$$

where C_t is the concentration of organisms surviving t(min) of treatment with a concentration, q, of lethal chemical compounds, C_q is the concentration of organisms surviving a fixed

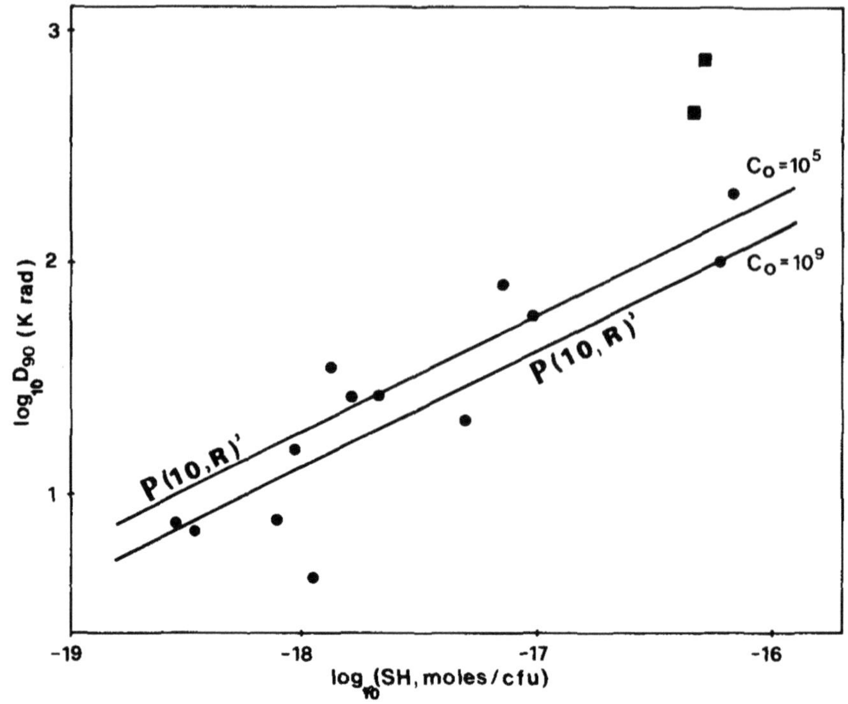

FIGURE 11. Relationship between the –SH level per cell and the first decimal reduction dose (comprehensive of the shoulder) of some bacteria, as reported by Bruce et al.,[168] together with the first decimal reduction dose P(10,R)′ as predicted by the general model using S′ = 0.25·S = surface –SH content/cell. (■) = *Micrococcus radiodurans*, (●) = other microorganisms.

time of treatment with a concentration, q, of lethal compounds, $Q = q^2$, and S is a sensitivity parameter specific to the microorganism being treated. Survivor curves from Equation 127 are inevitably concave (Figure 6); those from Equation 128 are convex, sigmoid, or concave, according to the lethal concentration of the chemical compound and specific microbial sensitivity (Figure 10).

A general model for disinfection has not been developed, but several observations seem to suggest that a more accurate description of survivor curves could be obtained by an equation of the following type:

$$C_{q,t} = C_o^{(1+Qt^2)^{-1}} \tag{129}$$

where:

$$Q = (Nm/1000)^2 \exp(-2E_d/RT) \tag{130}$$

m being the molarity of the lethal chemical compound used, N is Avogadro's number, and E_d is a measure of the specific sensitivity of the microorganism tested.

D. Observations

The general model developed seems to meet the fundamental requirements of a theory: it supplies a general mechanism of cell-environment interaction, based on random collisions between lighter molecules with sufficiently high energy and microorganisms; it describes most known experimental observations; it possesses a quite high predictive power; it suggests further improvements. In addition it can be used to describe microbial inactivation kinetics by heat, radiations, and chemical compounds, whatever the shape of the survivor curve; a single parameter (E_d), obtained simply from a single triplet of experimental data, can allow the description of inactivation kinetics, whatever the temperature and the water content of the medium; it supplies an exact explanation of the meaning of Q_{10} and z; changes of Q_{10} and z values following changes of water content of the medium can be easily computed; it explains radiation inactivation kinetics on the basis of microbial content of a defined chemical group, viz., –SH; it supplies a provisional theory of chemical inactivation kinetics, suggesting (Equations 129 and 130) a promising approach to a comprehensive theory of the disinfection process. Further, microbial growth kinetics can be described on the same basis.[155] More recently, the model has been used to fit human embryo and human tumor growth kinetics, linking the specific rate of growth to quantitatively defined molecules in the surface structure of metazoan cells.[170]

E. Practical Consequences of the Model

As far as the process of thermal sterilization is concerned, high temperatures must be preferred, since by increasing treatment temperature the probability of tailing decreases and microbial inactivation kinetics approaches exponential type behavior. A high water content in the substrate to be sterilized must be preferred, since it implies eventually higher temperature of treatment. Sterilization by ionizing radiations cannot be expected to be straightforward, even at very high doses, because of the high content of –SH groups in microbial surface structures, which probably operate by scavenging radiation-induced radicals (indirect effect) whose targets are expected to be enzymatic complexes with the ability to repair intracellular (mostly nucleic acids) radiation damage. A great deal of work must be done on application of the model to the disinfection process.

VI. CONCLUDING REMARKS

Lazzaro Spallanzani (1729—1799) proved in 1765 that microorganisms were killed by heat. Louis Pasteur (1822—1895) improved our knowledge of this phenomenon and promoted its practical application. Afterwards, experimental observations of microbial inactivation by physical and chemical agents increased in quality and number. Perhaps unfortunately, a first-order kinetics theory of microbial death was developed, having the attractive features that it was linked to chemical kinetics, it was seemingly valid for all lethal agents, and it was easy to treat mathematically. Later, other theories were developed, including the target theory.

At present, in an attempt to recognize in the cobweb of experimental observations and associated theories, only the events really relevant to understanding facts and mechanisms, I am inclined to believe that:

1. The theory of microbial exponential inactivation kinetics can be regarded as fundamentally wrong since it does not explain a large quantity of experimental data and it does not supply any insight into the mechanisms of microbial death.
2. The most relevant contribution to the understanding of the mechanisms of microbial death was made by Ball and Olson,[7] who suggested that "the cause of death must be outside the cell. It must be in the medium", and that at the microbial level statistical concepts break down and must be replaced "by energy considerations involving molecules in the discrete sense".
3. The general theory based on a random process of collision between "discrete" units and microorganisms as a fundamental mechanism of cell-environment interaction is able to accommodate all relevant experimental observations, is amenable to modifi-

cations, and can be regarded as a useful approach to the understanding of the mechanisms of death.

REFERENCES

1. **Schmidt, C. F.,** Thermal resistance of microorganisms, in *Antiseptics, Disinfectants, Fungicides and Sterilization,* Reddish, G. F., Ed., Lea & Febiger, Philadelphia, 1954, 720.
2. **Madsen, T. and Nyman, M.,** Zur theorie der desinfection, *Z. Hyg. Infekt.,* 57, 388, 1907.
3. **Chick, H.,** An investigation of the laws of disinfection, *J. Hyg.,* 8, 92, 1908.
4. **Chick, H.,** The process of disinfection by chemical agencies and hot water, *J. Hyg.,* 10, 237, 1910.
5. **Bigelow, W. D. and Esty, J. R.,** The thermal death point in relation to typical thermophilic organisms, *J. Infect. Dis.,* 27, 602, 1920.
6. **Esty, J. R. and Meyer, K. F.,** The heat resistance of the spores of *Bacillus botulinus* and allied anaerobes, *J. Infect. Dis.,* 31, 650, 1922.
7. **Ball, C. O. and Olson, F. C. W.,** *Sterilization in Food Technology,* McGraw-Hill, New York, 1957, 179.
8. **Stumbo, C. R.,** *Thermobacteriology in Food Processing,* Academic Press, New York, 1973, 70.
9. **Pflug, I. J. and Schmidt, C. F.,** Thermal destruction of microorganisms, in *Disinfection, Sterilization and Preservation,* Lawrence, C. A. and Block, S. S., Eds., Lea & Febiger, Philadelphia, 1968, 63.
10. **Esselen, W. B. and Pflug, I. J.,** Thermal resistance of PA 3679 spores in vegetables in the temperature range of 250—290°F, *Food Technol.,* 10, 557, 1956.
11. **Kaplan, C.,** The heat inactivation of Vaccinia virus, *J. Gen. Microbiol.,* 18, 58, 1958.
12. **Amaha, M.,** Factors affecting heat-destruction of bacterial spores, in *Proc. 4th Int. Cong. on Canned Foods,* Berlin, 1961, 13.
13. **Edwards, J. L., Busta, F. F., and Speck, M. L.,** Thermal inactivation characteristics of *Bacillus subtilis* spores at ultrahigh temperatures, *Appl. Microbiol.,* 13, 858, 1965.
14. **Casolari, A. and Taffurelli, M.,** unpublished data, 1980.
15. **Leffler, J. E.,** Enthalpy-entropy relation and its implication for organic chemistry, *J. Org. Chem.,* 20, 1202, 1955.
16. **Blackadder, D. A. and Hinsellwood, C. N.,** Isokinetic relationship in organic chemistry, *J. Chem. Soc.,* 2720, 2278, 1958.
17. **Rosenberg, B., Kemeny, G., Switzer, R. C., and Hamilton, T. C.,** Quantitative evidence for protein denaturation as the cause of thermal death, *Nature (London),* 232, 471, 1971.
18. **Bancks, B. E. C., Damjanovic, V., and Vernon, C. A.,** The so-called thermodynamic compensation law and thermal death, *Nature (London),* 240, 147, 1972.
19. **Casolari, A.,** Uncertainties as to the kinetics of heat inactivation of microorganisms, in *Proc. Int. Meeting on Food Microbiology and Technology,* Jarvis, B., Christian, J. N. B., and Michener, H. D., Eds., Medicina Viva, Parma, Italy, 1979, 231.
20. **Tischer, R. G. and Hurwicz, H.,** Thermal characteristics of bacterial populations, *Food Res.,* 19, 80, 1954.
21. **Charm, S. E.,** The kinetics of bacterial inactivation by heat, *Food Technol.,* 12, 4, 1958.
22. **Shechmeister, I. L.,** Sterilization by ultraviolet radiation, in *Disinfection, Sterilization and Preservation,* Lawrence, C. A. and Block, S. S., Eds., Lea & Febiger, Philadelphia, 1968, 761.
23. **Morris, E. J. and Darlow, H. M.,** Inactivation of virus, in *Inhibition and Destruction of the Microbial Cell,* Hugo, W. B., Ed., Academic Press, New York, 1971, 687.
24. **Russell, A. D.,** The destruction of bacterial spores, in *Inhibition and Destruction of the Microbial Cell,* Hugo, W. B., Ed., Academic Press, New York, 1971, 451.
25. **Gola, S.,** Studio della resistenza microbica alle radiazioni ultraviolette, *Ind. Cons.,* 57, 26, 1982.
26. **Fano, U.,** Principles of radiobiological physics, in *Radiation Biology,* Hollaender, A., Ed., McGraw-Hill, New York, 1954, 1.
27. **Zirkle, R. E.,** The radiobiological importance of linear energy transfer, in *Radiation Biology,* Hollaender, A., Ed., McGraw-Hill, New York, 1954, 315.
28. **Rayman, M. M. and Byrne, A. F.,** Action of ionizing radiation on microorganisms, in Radiation Preservation of Foods, U. S. Army Quartermaster Corps, U. S. Government Printing Office, Washington, D.C., 1957, 208.
29. **Dale, W. M.,** Basic radiation biochemistry, in *Radiation Biology,* Hollaender, A., Ed., McGraw-Hill, New York, 1954, 255.

30. **Lea, O. E.,** *Action of Radiations on Living Cells*, Cambridge University Press, London, 1956.
31. **Silverman, G. J. and Sinskey, T. J.,** The destruction of microorganisms by ionizing radiation, in *Disinfection, Sterilization and Preservation*, Lawrence, C. A. and Block, S. S., Eds., Lea & Febiger, Philadelphia, 1968, 741.
32. **Casolari, A.,** Sensibilità dei batteri alle radiazioni ionizzanti, *Ind. Cons.*, 42, 189, 1967.
33. **Lucisano, A., Casolari, A., and Cultrera, R.,** Relative sensitivities of mercaptopyrimidines and mercaptopurines to –SH reagents and gamma radiation, *Ann. Chim.*, 61, 527, 1971.
34. **Casolari, A., Campanini, M., and Cicognani, G.,** Sensibilità alle radiazioni in spore pretrattate termicamente, *Ind. Cons.*, 44, 199, 1969.
35. **Cultrera, R., Casolari, A., and Campanini, M.,** Radioresistenza e attivazione sporale da periodato, *Ind. Cons.*, 44, 204, 1969.
36. **Goldblith, S. A.,** The inhibition and destruction of the microbial cell by radiations, in *Inhibition and Destruction of the Microbial Cell*, Hugo, W. B., Ed., Academic Press, New York, 1971, 285.
37. **Hoffman, R. K. and Warshowsky, B.,** Beta-propiolactone vapor as a disinfectant, *Appl. Microbiol.*, 6, 358, 1958.
38. **Rubbo, G. D., Gardner, G. F., and Webb, R. L.,** Biocidal activity of glutaraldehyde and related compounds, *J. Appl. Bacteriol.*, 30, 78, 1967.
39. **Swartling, P. and Lindgren, B.,** The sterilizing effect against *Bacillus subtilis* spores of hydrogen peroxide at different temperatures and concentrations, *J. Dairy Res.*, 35, 423, 1968.
40. **Levine, M.,** Spores as reagents for studies on chemical disinfection, *Bacteriol. Rev.*, 16, 117, 1952.
41. **Toledo, R. T., Escher, F. E., and Ayres, J. C.,** Sporicidal properties of hydrogen peroxide against food spoilage organisms, *Appl. Microbiol.*, 26, 592, 1973.
42. **Cerf, O. and Metro, F.,** Tailing of survival curves of *Bacillus licheniformis* spores treated with hydrogen peroxide, *J. Appl. Bacteriol.*, 42, 405, 1977.
43. **Gelinas, P., Goulet, J., Tastayre, G. M., and Picard, G. A.,** Effect of temperature and contact time on the activity of eight disinfectants. A classification, *J. Food Protect.*, 47, 841, 1984.
44. **Kostenbauder, H. B.,** Physical factors influencing the activity of antimicrobial agents, in *Sterilization, Disinfection and Preservation*, Lawrence, C. A. and Block, S. S., Eds., Lea & Febiger, Philadelphia, 1968, 44.
45. **Jones, L. A., Hoffman, R. K., and Phillips, C. R.,** Sporicidal activity of peracetic acid and beta-propiolactone at subzero temperatures, *Appl. Microbiol.*, 15, 357, 1967.
46. **Bernarde, M. A., Snow, W. B., and Olivieri, V. P.,** Chlorine dioxide disinfection temperature effects, *J. Appl. Bacteriol.*, 30, 159, 1967.
47. **Sykes, G.,** The sporicidal properties of chemical disinfectants, *J. Appl. Bacteriol.*, 33, 147, 1970.
48. **Hoffman, R. K.,** Toxic gases, in *Inhibition and Destruction of the Microbial Cell*, Hugo, W. B., Ed., Academic Press, London, 1971, 226.
49. **Ito, K. A., Denny, C. B., Brown, C. K., and Seeger, M. L.,** Resistance of bacterial spores to hydrogen peroxide, *Food Technol.*, 27, 58, 1973.
50. **Brown, M. R. W. and Melling, J.,** Inhibition and destruction of microorganisms by heat, in *Inhibition and Destruction of the Microbial Cell*, Hugo, W. B., Ed., Academic Press, New York, 1971, 1.
51. **Sugiyama, H.,** Studies on factors affecting the heat resistance of spores of *Clostridium botulinum*, *Food Res.*, 16, 81, 1951.
52. **Yokoya, F. and York, G. K.,** Effect of several environmental conditions on the thermal death rate of endospores of aerobic, thermophilic bacteria, *Appl. Microbiol.*, 13, 993, 1965.
53. **Hurst, A.,** Bacterial injury: a review, *Can. J. Microbiol.*, 23, 935, 1977.
54. **Camper, A. K. and McFeters, L.,** Chlorine injury and enumeration of waterborn coliform bacteria, *Appl. Environ. Microbiol.*, 37, 633, 1979.
55. **Foegeding, P. M. and Busta, F. F.,** Bacterial spore injury. An update, *J. Food Protect.*, 44, 776, 1981.
56. **Moats, W. A., Dabbah, R., and Edwards, V. M.,** Interpretation of nonlogarithmic survivor curves of heated bacteria, *J. Food Sci.*, 36, 523, 1971.
57. **Moats, W. A.,** Kinetics of thermal death of bacteria, *J. Bacteriol.*, 105, 165, 1971.
58. **Condon, E. U. and Terrill, H. M.,** Quantum phenomena in the biological action of X-rays, *J. Cancer Res.*, 11, 324, 1927.
59. **Curie, P.,** Sur l'étude des curbes de probabilité relative à l'action des rayons X sur les bacilles, *Comptes Rendus*, 188, 202, 1929.
60. **Wyckoff, R. W. G. and Rivers, T. M.,** The effect of cathode rays upon certain bacteria, *J. Exp. Med.*, 51, 921, 1949.
61. **Atwood, K. C. and Norman, A.,** On the interpretation of multihit survival curves, *Proc. Natl. Acad. Sci. U.S.A.*, 35, 596, 1949.
62. **Alper, T., Gillies, N. E., and Elkind, M. M.,** The sigmoid survival curve in radiobiology, *Nature (London)*, 186, 1062, 1960.

63. **Peled, O. N., Salvadori, A., Peled, U. N., and Kidby, D. K.,** Death of microbial cells: rate constant calculations, *J. Bacteriol.*, 129, 1648, 1977.
64. **Alderton, G. and Snell, N.,** Chemical states of bacterial spores: heat resistance and its kinetics at intermediate water activity, *Appl. Microbiol.*, 19, 565, 1970.
65. **King, A. D., Bayne, H. G., and Alderton, G.,** Nonlogarithmic death rate calculations for *Byssochlamys fulva* and other microorganisms, *Appl. Environ. Microbiol.*, 37, 596, 1979.
66. **Rahn, O.,** Physical methods of sterilization of microorganisms, *Bacteriol. Rev.*, 9, 1, 1945.
67. **Mackenzie, D. W.,** Effect of relative humidity on survival of *Candida albicans* and other yeasts, *Appl. Microbiol.*, 22, 678, 1971.
68. **Cox, C. S.,** Inactivation of some microorganisms subjected to a variety of stresses, *Appl. Environ. Microbiol.*, 31, 836, 1976.
69. **Riley, R. L. and Kaufman, J. E.,** Effect of relative humidity on the inactivation of airborne *Serratia marcescens* by UV radiation, *Appl. Microbiol.*, 23, 1113, 1972.
70. **Gilbert, G. L., Gambill, V. M., Spiner, D. R., Hoffman, R. K., and Phillips, C. R.,** Effect of moisture on ethylene oxide sterilization, *Appl. Microbiol.*, 12, 496, 1964.
71. **Juven, B. J., Ben-Shalom, N., and Weisslowicz, H.,** Identification of chemical constituents of tomato juice which affect the heat resistance of *Lactobacillus fermentum*, *J. Appl. Bacteriol.*, 54, 335, 1983.
72. **Verrips, C. T., Glas, R., and Kwast, R. H.,** Heat resistance of *Klebsiella pneumoniae* in media with various sucrose concentrations, *Eur. J. Appl. Microbiol. Biotechnol.*, 8, 299, 1979.
73. **Walker, G. C. and Harmon, L. G.,** Thermal resistance of *Staphylococcus aureus* in milk, whey and phosphate buffer, *Appl. Microbiol.*, 14, 584, 1966.
74. **Casolari, A. and Campanini, M.,** Resistenza termica in Lactobacillaceae, *Ind. Cons.*, 48, 140, 1973.
75. **Campanini, M., Mussato, G., Barbuti, S., and Casolari, A.,** Resistenza termica di streptococchi isolati da mortadelle alterate, *Ind. Cons.*, 59, 298, 1984.
76. **Casolari, A. and Cagna, D.,** Sull'inattivazione termica del P.A. 3679 nelle conserve di tonno, *Ind. Cons.*, 48, 69, 1973.
77. **Walker, H. W., Matches, J. R., and Ayres, J. C.,** Chemical composition and heat resistance of some aerobic bacterial spores, *J. Bacteriol.*, 82, 960, 1961.
78. **Fox, K. and Pflug, I. J.,** Effect of temperature and gas velocity on the dry-heat destruction rate of bacterial spores, *Appl. Microbiol.*, 16, 343, 1968.
79. **Farmiloe, F. J., Cornford, S. J., Coppock, J. B. M., and Ingram, M.,** The survival of *Bacillus subtilis* spores in the baking of bread, *J. Sci. Food Agric.*, 5, 292, 1954.
80. **Gibriel, A. Y. and Abd-El-Al, A. T. H.,** Measurement of heat resistance parameters for spores isolated from canned products, *J. Appl. Bacteriol.*, 36, 321, 1973.
81. **Roberts, T. A., Gilbert, R. J., and Ingram, M.,** The heat resistance of anaerobic spores, *J. Food Technol.*, 1, 227, 1966.
82. **Casolari, A.,** Non-logarithmic behaviour of heat inactivation curves of PA 3679 spores, in *Proc. 4th Int. Cong. Food Sci. Technol.*, Vol. 3, 1974, 86.
83. **Corry, J. E. L.,** The effect of sugars and polyols on the heat resistance of osmophilic yeasts, *J. Appl. Bacteriol.*, 40, 269, 1976.
84. **Christophersen, J. and Precht, H.,** Untersuchungen zum problem der hitzeresistenz. II. Untersuchungen an hefezellen, *Biol. Zentralbl.*, 71, 585, 1952.
85. **Morris, E. O.,** Yeast growth, in *The Chemistry and Biology of Yeasts*, Cook, A. H., Ed., Academic Press, New York, 1958, 251.
86. **Casolari, A. and Castelvetri, F.,** Studio dei lieviti osmofili, *Ind. Cons.*, 52, 105, 1977.
87. **Lubieniecki von Schelhorn, M.,** Influence of relative humidity on the thermal resistance of several kinds of spores of molds, *Acta Aliment.*, 2, 163, 1973.
88. **White, A., Berman, S., and Lowenthal, J. P.,** Inactivated eastern encephalomyelitis vaccines prepared in monolayer and concentrated suspension chick embryo cultures, *Appl. Microbiol.*, 22, 909, 1971.
89. **Koch, G.,** Influence of assay conditions on infectivity of heated poliovirus, *Virology*, 12, 601, 1960.
90. **Hiatt, C. W.,** Kinetics of the inactivation of viruses, *Bacteriol. Rev.*, 28, 150, 1964.
91. **Wilkowske, H. H., Nelson, F. E., and Parmelee, C. E.,** Heat inactivation of bacteriophages active against lactic streptococci, *Appl. Microbiol.*, 2, 250, 1954.
92. **Sullivan, R., Tierney, J. T., Larkin, E. P., Read, R. B., and Peeler, J. T.,** Thermal resistance of certain oncogenic viruses suspended in milk and milk products, *Appl. Microbiol.*, 22, 315, 1971.
93. **Jamlang, E. M., Bartlett, M. L., and Snyder, H. E.,** Effect of pH, protein concentration and ionic strength on heat inactivation of *Staphylococcus* enterotoxin B, *Appl. Microbiol.*, 22, 1034, 1971.
94. **Woodburn, M. J., Somers, E., Rodriguez, J., and Schantz, E. J.,** Heat inactivation rates of botulinum toxins A, B, E and F in some foods and buffers, *J. Food Sci.*, 44, 1658, 1979.
95. **Babajimopoulos, M. and Mikolajcik, E. M.,** Heat inactivation and reactivation of alpha toxin from *Clostridium perfringens*, *J. Food Protect.*, 44, 899, 1981.

96. **Patel, T. R., Bartlett, F. M., and Hamid, J.,** Extracellular heat resistant proteases of psychrotrophic *Pseudomonas*, *J. Food Protect.*, 46, 90, 1983.
97. **Sadoff, H. L.,** Heat resistance of spore enzymes, *J. Appl. Bacteriol.*, 33, 130, 1970.
98. **Trujillo, R. and David, T. J.,** Sporostatic and sporicidal properties of aqueous formaldehyde, *Appl. Microbiol.*, 23, 618, 1972.
99. **Russell, A. D. and Loosemore, M.,** Effect of phenol on *Bacillus subtilis* spores at elevated temperatures, *Appl. Microbiol.*, 12, 403, 1964.
100. **Eubanks, V. L. and Beuchat, L. R.,** Combined effects of antioxidants and temperature on survival of *Saccharomyces cerevisiae*, *J. Food Protect.*, 46, 29, 1983.
101. **Bayliss, C. E. and Waites, W. M.,** The combined effect of hydrogen peroxide and ultraviolet irradiation on bacterial spores, *J. Appl. Bacteriol.*, 47, 263, 1979.
102. **Burgos, J., Ordonez, J. A., and Sala, F.,** Effects of ultrasonic waves on the heat resistance of *Bacillus cereus* and *Bacillus licheniformis* spores, *Appl. Microbiol.*, 24, 497, 1972.
103. **Parkinson, A. J., Muchmore, H. G., Scott, E. N., and Scott, L. V.,** Survival of human parainfluenza viruses in the South Polar environment, *Appl. Environ. Microbiol.*, 46, 901, 1983.
104. **Maxcy, R. B., Tawari, N. P., and Soprey, P. R.,** Changes in *E. coli* associated with acquired tolerance for quaternary ammonium compounds, *Appl. Microbiol.*, 22, 229, 1971.
105. **Michener, H. D. and Elliot, R. P.,** Microbiological conditions affecting frozen food quality, in *Quality and Stability of Frozen Foods*, Van Ardsel, W. B., Copley, M. J., and Olson, R. L., Eds., John Wiley & Sons, New York, 1969, 43.
106. **Georgala, D. L. and Hurst, A.,** The survival of food poisoning bacteria in frozen foods, *J. Appl. Bacteriol.*, 26, 346, 1963.
107. **Hess, G. E.,** Effect of oxygen on aerosolized *Serratia marcescens*, *Appl. Microbiol.*, 13, 781, 1965.
108. **Burleson, G. R., Murray, T. M., and Pollard, M.,** Inactivation of viruses and bacteria by ozone, with and without sonication, *Appl. Microbiol.*, 29, 340, 1975.
109. **Alvarez, M. E. and O'Brien, R. T.,** Mechanism of inactivation of poliovirus by chlorine dioxide and iodine, *Appl. Environ. Microbiol.*, 44, 1064, 1982.
110. **Cove, J. H. and Holland, K. T.,** The effect of benzoyl peroxide on cutaneous microorganisms in vitro, *J. Appl. Bacteriol.*, 54, 379, 1983.
111. **Grabow, W. O. K., Gauss-Muller, V., Prozesky, O. W., and Deinhardt, F.,** Inactivation of hepatitis A virus and indicator organisms in water by free chlorine residuals, *Appl. Environ. Microbiol.*, 46, 619, 1983.
112. **Ridgway, H. F. and Olson, B. H.,** Chlorine resistance patterns of bacteria from two drinking water distribution systems, *Appl. Environ. Microbiol.*, 44, 972, 1982.
113. **Greenspan, F. P. and MacKellar, D. G.,** The application of peracetic acid germicidal washes to mold control of tomatoes, *Food Technol.*, 5, 95, 1951.
114. **Thomas, S. and Russell, A. D.,** Studies on the mechanism of the sporicidal action of glutaraldehyde, *J. Appl. Bacteriol.*, 37, 83, 1974.
115. **Futter, B. V. and Richardson, G.,** Inactivation of bacterial spores by visible radiations, *J. Appl. Bacteriol.*, 30, 347, 1967.
116. **Keller, L. C. and Maxcy, R. B.,** Effect of physiological age on radiation resistance of some bacteria that are high radiation resistant, *Appl. Environ. Microbiol.*, 47, 915, 1984.
117. **Moseley, B. E. B. and Laser, H.,** Similarity of repair of ionizing and ultraviolet radiation damage in *Micrococcus radiodurans*, *Nature (London)*, 206, 377, 1965.
118. **Cerny, G.,** Entkeimen von packstoffen beim aseptischen abpacken. 2.Mitt.: Untersuchungen zur keimabtotenden wirkung von UVC strahlen, *Verpack.-Rundsch.*, 28, 77, 1977.
119. **Lippert, K. D.,** Abtotung von schimmelconidien in vorgeformtem verpackungsmaterial durch UV-strahlung, *Verpack.-Rundsch.*, 7, 51, 1979.
120. **Abshire, R. L., Bain, B., and Williams, T.,** Resistance and recovery studies on ultraviolet irradiated spores of *Bacillus pumilus*, *Appl. Environ. Microbiol.*, 39, 695, 1980.
121. **Denny, C. B., Bohrer, W. C., Perkins, W. E., and Townsend, C. T.,** Destruction of *Clostridium botulinum* by ionizing radiation, *Food Res.*, 24, 44, 1958.
122. **Grecz, N.,** Biophysical aspects of Clostridia, *J. Appl. Bacteriol.*, 28, 17, 1965.
123. **Anellis, A., Grecz, N., and Berkowitz, D.,** Survival of *Clostridium botulinum* spores, *Appl. Microbiol.*, 13, 397, 1965.
124. **Wheaton, E. and Pratt, G. B.,** Radiation survival curves of *Clostridium botulinum* spores, *J. Food Sci.*, 27, 327, 1962.
125. **Masokhina-Porsniakova, N. N. and Ladukina, G. V.,** The effect of ionizing radiation on *Clostridium botulinum* spores, in *Microbiological Problems in Food Preservation by Irradiation*, IAEA Panel Proc. Ser., International Atomic Energy Agency, Vienna, 1967, 27.
126. **Morgan, B. H. and Reed, J. M.,** Resistance of bacterial spores to gamma irradiation, *Food Res.*, 19, 357, 1954.

127. **Woese, C. R.,** Comparison of the X-ray sensitivity of bacterial spores, *J. Bacteriol.*, 75, 5, 1958.
128. **Woese, C. R.,** Further studies on the ionizing radiation inactivation of bacterial spores, *J. Bacteriol.*, 77, 38, 1959.
129. **Roberts, T. A. and Ingram, M.,** The resistance of spores of *Clostridium botulinum* type E to heat and radiation, *J. Appl. Bacteriol.*, 28, 125, 1965.
130. **Dean, C. J. and Alexander, P.,** Sensitization of radio-resistant bacteria to X-rays by iodoacetamide, *Nature (London)*, 196, 1324, 1962.
131. **Quinn, D. J., Anderson, A. W., and Dyer, J. F.,** The inactivation of infection and intoxication microorganisms by irradiation in seafood, in *Microbiological Problems in Food Preservation by Irradiation*, IAEA Panel Proc. Ser., International Atomic Energy Agency, Vienna, 1967, 1.
132. **Johnson, C. D.,** X-rays inactivation of foot-and-mouth disease virus and vesicular stomatitis virus in aqueous media, in *Microbiological Problems in Food Preservation by Irradiation*, IAEA Panel Proc. Ser., International Atomic Energy Agency, Vienna, 1967, 65.
133. **DeGuzman, A., Fields, M. L., Humbert, R. D., and Kazanas, N.,** Sporulation and heat resistance of *Bacillus stearothermophilus* spores produced in chemically defined medium, *J. Bacteriol.*, 110, 775, 1972.
134. **Cerf, O., Berry, J.-L., Riottot, M., and Bouvet, Y.,** Un appareil simple pour la mesure de l'activité de solutions désinfectantes ou sterilizantes à action rapide, et son application à la mesure de l'action de l'hypochlorite de sodium sur les spores bacteriennes, *Pathol. Biol.*, 21, 889, 1973.
135. **Han, Y. W., Zhang, H. I., and Krochta, J. M.,** Death rate of bacterial spores: mathematical models, *Can. J. Microbiol.*, 22, 295, 1976.
136. **Sharpe, K. and Bektash, R. M.,** Heterogeneity and the modelling of bacterial spore death: the case of continuously decreasing death rate, *Can. J. Microbiol.*, 23, 1501, 1977.
137. **Bond, W. W., Favero, M. S., Petersen, N. J., and Marshall, J. H.,** Dry-heat inactivation kinetics of naturally occurring spore populations, *Appl. Microbiol.*, 20, 573, 1970.
138. **Vas, K. and Prostz, G.,** Observations on the heat destruction of spores of *Bacillus cereus*, *J. Appl. Bacteriol.*, 20, 431, 1957.
139. **Fernelius, A. L., Wilkes, C. E., DeArmon, I. A., and Lincoln, R. E.,** A probit method to interpret thermal inactivation of bacterial spores, *J. Bacteriol.*, 75, 300, 1956.
140. **Levinson, H. S. and Hyatt, M. T.,** Distribution and correlation of events during thermal inactivation of *Bacillus megaterium*, *J. Bacteriol.*, 108, 111, 1971.
141. **Alderton, G., Thompson, P. A., and Snell, N.,** Heat adaptation and ion exchange in *Bacillus megaterium* spores, *Science*, 143, 141, 1964.
142. **Komemushi, S. and Terui, G.,** On the change of death rate constant of bacterial spores in the course of heat sterilization, *J. Ferm. Technol.*, 45, 764, 1967.
143. **Han, Y. W.,** Death rate of bacterial spores: nonlinear survivor curves, *Can. J. Microbiol.*, 21, 1464, 1975.
144. **Brannen, J. P.,** A rational model for thermal sterilization of microorganisms, *Math. Biosci.*, 2, 165, 1968.
145. **Migaky, H. and McCulloch, E. C.,** Survivor curves of bacteria exposed to surface-active agents, *J. Bacteriol.*, 58, 161, 1949.
146. **Stumbo, C. R., Murphy, J. R., and Cochran, J.,** Nature of thermal death time curves for *P.A. 3679* and *Clostridium botulinum*, *Food Technol.*, 4, 321, 1950.
147. **Reed, J. M., Bohrer, C. W., and Cameron, E. J.,** Spore destruction rate studies on organisms of significance in the processing of canned foods, *Food Res.*, 16, 383, 1951.
148. **Pflug, J. I. and Esselen, W. B.,** Development and application of an apparatus for study of thermal resistance of bacterial spores and thiamine at temperatures above 250°F, *Food Technol.*, 7, 237, 1953.
149. **Kempe, L. L., Graikowski, J. T., and Gillies, R. A.,** Gamma ray sterilization of canned meat previously inoculated with anaerobic bacterial spores, *Appl. Microbiol.*, 2, 230, 1954.
150. **Gola, S. and Casolari, A.,** Antimicrobial activity of nitrite-dependent compounds, in *Food Microbiology and Technology*, Jarvis, B., Christian, J. H. B., and Michener, D. H., Eds., Medicina Viva, Parma, 1979, 267.
151. **Greenberg, R. A.,** Nitrite in the control of *Clostridium botulinum*, *Proc. Am. Meat Inst. Found. Meat Ind. Res. Conf.*, 1972, 25.
152. **Crowther, J. S., Holbrook, R., Baird-Parker, A. C., and Austin, B. L.,** Role of nitrite and ascorbate in the microbiological safety of vacuum packed sliced bacon, *Proc. 2nd Int. Symp. on Nitrite in Meat Products*, Krol, E. and Tinbergen, B. J., Eds., Pudoc, Wageningen, 1976, 13.
153. **Roberts, T. A. and Ingram, M.,** Nitrite and nitrate in the control of *Clostridium botulinum* in cured meats, in *Proc. 2nd Int. Symp. on Nitrite in Meats*, Krol, B. and Tinbergen, B. J., Eds., Pudoc, Wageningen, 1976, 29.
154. **Casolari, A. and Medeot, L.,** unpublished data, 1978.
155. **Casolari, A.,** A model describing microbial inactivation and growth kinetics, *J. Theor. Biol.*, 88, 1, 1981.
156. **Jones, M. C.,** The temperature dependence of the lethal rate in sterilization calculations, *J. Food Technol.*, 3, 31, 1968.

157. **Cowell, N. D.,** Methods of thermal process evaluation, *J. Food Technol.*, 3, 303, 1968.
158. **Cerf, O.,** Tailing of survival curves of bacterial spores: a review, *J. Appl. Bacteriol.*, 42, 1, 1977.
159. **Murrell, W. G. and Scott, W. J.,** Heat resistance of bacterial spores at various water activities, *Nature (London)*, 179, 481, 1957.
160. **Pflug, I. J.,** Thermal resistance of microorganisms to dry heat: design of apparatus, operational problems and preliminary results, *Food Technol.*, 14, 483, 1960.
161. **Bruch, C. W.,** Dry heat sterilization for planetary-impacting spacecraft, in *Proc. Natl. Conf. on Spacecraft Sterilization Technology*, NASA SP-108, National Aeronautics and Space Administration, Washington, D.C., 1965.
162. **Harnulv, B. G. and Snygg, B. G.,** Heat resistance of *Bacillus subtilis* spores at various water activities, *J. Appl. Bacteriol.*, 35, 615, 1972.
163. **Russell, A. D. and Harries, D.,** Damage to *Escherichia coli* on exposure to moist heat, *Appl. Microbiol.*, 16, 1349, 1968.
164. **Casolari, A., Campanini, M., and Cicognani, G.,** Sulla diversa sensibilità allo iodio, al bromo e al cloro in spore dormienti, attivate e danneggiate termicamente, *Ind. Cons.*, 43, 211, 1968.
165. **Fowers, R. S. and Adams, D. M.,** Spores membrane(s) as the site of damage within heated *Clostridium perfringens* spores, *J. Bacteriol.*, 125, 429, 1976.
166. **Kroll, R. G. and Anagnostopoulos, G. D.,** Potassium leakage as a lethality index of phenol and the effect of solute and water activity, *J. Appl. Bacteriol.*, 50, 139, 1981.
167. **Degre, R. and Silvestre, M.,** Effect of butylated hydroxianisole on the cytoplasmic membrane of *Staphylococcus aureus*, *J. Food Protect.*, 46, 206, 1983.
168. **Bruce, A. K., Sansone, P. A., and MacVittie, T. J.,** Radioresistance of bacteria as a function of *p*-hydroxymercuribenzoate binding, *Radiat. Res.*, 38, 95, 1969.
169. **Moseley, B. E. B. and Laser, H.,** Repair of X-ray damage in *Micrococcus radiodurans*, *Proc. R. Soc. London Ser. B*, 162, 210, 1965.
170. **Casolari, A.,** Human embryo and human tumour growth kinetics, manuscript in preparation, 1987.

Chapter 8

THE KINETICS OF CHANGE IN BACTERIAL SPORE GERMINATION

Gerald M. Lefebvre and Raymond Leblanc

TABLE OF CONTENTS

I. INTRODUCTION

The dormant bacterial spore has been described as both an end point and a beginning in biology.[1] Through a sequence of genetic, biochemical, morphological, and physiological events, the vegetative cell gives rise to the spore in a process called sporogenesis. On the other hand, the spore responds to appropriate environmental stimuli by leaving the dormant state, undergoing rather drastic morphological and biochemical changes, eventually synthesizing those substances, and fashioning those structures which permit resumption of vegetative existence. This transformation from dormant state to vegetative form is called germination. The processes of sporogenesis and germination are examples of intracellular differentiation induced by appropriate environmental stimuli.

A considerable body of information exists on the morphological, biochemical, and physicochemical changes which occur in the bacterial spore in its transition to a vegetative form.[2,3] These events can be localized in one or another of an ordered sequence of four distinct and experimentally isolatable phases, namely activation, triggering, initiation, and outgrowth.[4] Activation describes the effects on a spore sample which has been subjected to various pretreatments resulting in changes in the rate and extent of germination, as well as in qualitative and quantitative modifications in the list of effective germinants.[5] Several of these activation techniques have been found to be reversible, while others are not. The spores of some strains of bacteria are very reluctant to germinate without some form of activation, while others will do so completely without pretreatment. Triggering is a direct consequence of the interaction of a spore with a germinant, and is an irreversible commitment to complete the degradative phase which follows. That a triggered spore does not require the continued presence of a germinant has been shown in a number of investigations.[6-9] During initiation, the phase of degradative events to which the triggered spore is committed, those properties which characterize the dormant state, i.e., a rigid structure, the absence of metabolic activity, and an augmented resistance to chemical and physical challenges, are lost. Many measurable changes occurring in this phase, including Ca^{2+} and dipicolinic acid release, loss of heat resistance, increased stainability, loss of refractility under phase contrast, and loss of absorbance by a suspension of spores, have traditionally been utilized to monitor germination. These have been shown to occur in a time-ordered sequence.[10-13] Given an appropriate environment, which may be different from that required for triggering, the spore initiates synthesis of those substances and structures leading to the formation of a vegetative bacillus. This last phase is called outgrowth by some authors.

A useful description of the process of germination might take the form of a time-ordered sequence of biochemical and biophysical events, together with their interdependencies, from which the various indices which characterize this process could be seen to arise. From such a basis one might eventually be able to identify various spores states, and to describe these, and the transitions linking them, in thermodynamic terms. In this context, then, it is from a mosaic of many different measurable events that the process of germination and its dependence on environmental conditions can find expression.

II. THE MEASURABLE EVENTS IN GERMINATION

A. Their Occurrence as a Time-Ordered Sequence

The observable changes characteristic of germination do not occur simultaneously in a single spore, but rather constitute a time-ordered sequence. This was first demonstrated in a conclusive way by Levinson and Hyatt[10] in *Bacillus megaterium* QM B1551. Utilizing 2 mM $HgCl_2$ to stabilize germination levels at various times after mixing of spores and germinants, the extent of completion of various characteristic changes could be compared. Their results, for spores prepared in two different sporulating media, and germinated under

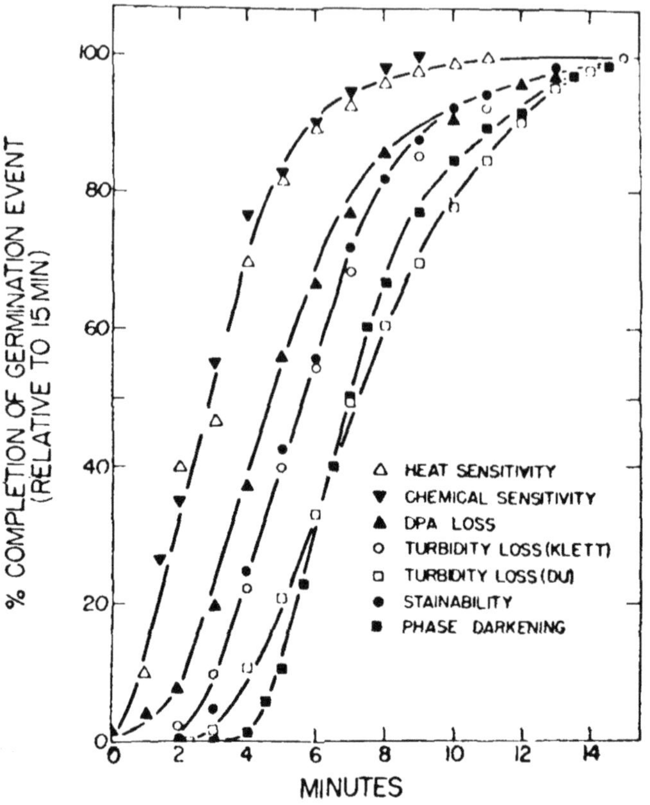

FIGURE 1. Kinetics of *Bacillus megaterium* spore germination events. Spores were produced on liver extract (LB) medium. Heat activated (60°C for 10 min) spores were germinated by incubating them at 30°C with a mixture of glucose (0.025 *M)* and L-alanine (0.001 *M)* in 0.05 *M* phosphate buffer (pH 7.0). (From Levinson, H. S. and Hyatt, M., *J. Bacteriol.*, 91, 1811, 1966. With permission.)

identical conditions, are reproduced in Figures 1 and 2. A certain degree of parallelism between the various response curves can be seen for spores prepared on the LB medium, whereas there are obviously two classes of response curves in A-K prepared spores. Germination on one event can evidently be different from this same process described on another, even within the same spore sample. As pointed out by these investigators in referring to Figure 1, " . . . after five minutes incubation, depending on the germination criterion, one might conclude that germination had progressed to 82% (heat or chemical sensitivity), 56% (DPA loss), 43% (stainability), 21% (absorbance loss), or 11% (phase darkening) of completion". Taking 50% completion as a reasonable indicator that one half of the spores in the sample have completed the change in that observable property, again with reference to the LB preparation, these characteristic times are 3.0 min (heat or chemical sensitivity), 4.7 min (DPA loss), 7.0 min (phase darkening), and 7.2 min (absorbance loss). Were perfect parallelism observed in each set of response curves, their order of occurrence could then be described in terms of the lag time of each curve. Moreoever, in that case a mathematical expression capable of describing the time evolution of the sample for one event could, duly corrected for differences in sample lag times, serve equally well for any other. One could then select a convenient event, and its time evolution would describe the kinetics of germination. In order to take into account differences in response curves, it will be necessary to more closely examine those elements likely to influence the shape of these curves. Any

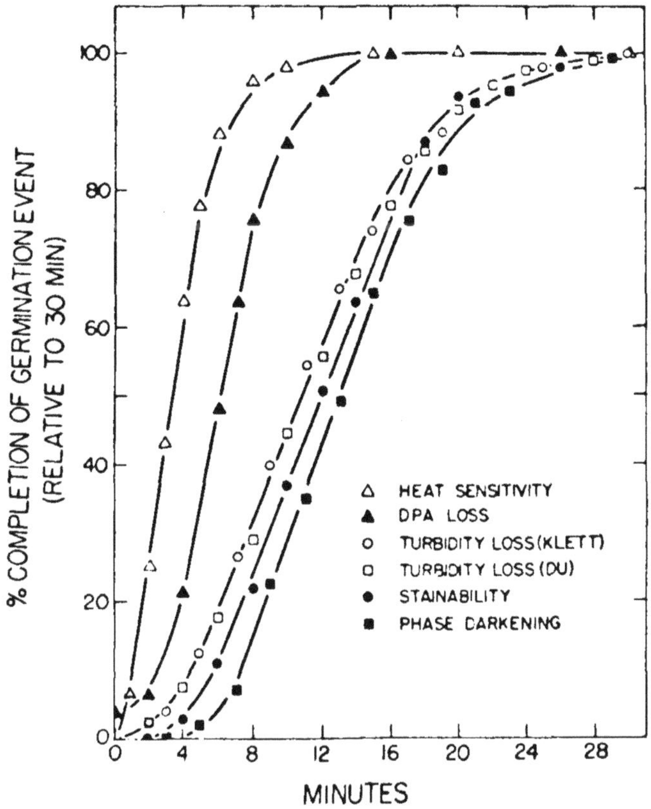

FIGURE 2. Kinetics of *Bacillus megaterium* spore germination events. Spores were produced on Arret-Kirshbaum (A-K) medium. Germination conditions were as in Figure 1. (From Levinson, H. S. and Hyatt, M., *J. Bacteriol.*, 91, 1811, 1966. With permission.)

observable property which undergoes change during the germination process can, in principle, serve to monitor this process. As will be illustrated in detail below, each observable event can be labeled with a unique coefficient ϵ. The change in this coefficient from its initial value ϵ_0 to its final value ϵ_∞ is the contribution of an individual spore to the change in the corresponding sample property.

B. Microlag and Microgermination

An early indication of the importance of the asynchrony in the spores of a sample undergoing germination in determining the shapes of the various response curves can be found in a study by Pulvertaft and Haynes.[14] A cinematographic record of darkening under phase contrast microscopy was made of a sample of germinating *B. cereus* T spores. This change was found to be virtually instantaneous in the individual spore when compared to the sample germination time. However, the interval required before this change occurred was highly variable among the spores of the sample. Some germinated shortly after addition of the germinant, while a few remained bright for most of the sample germination time before rapidly darkening. Therefore, the spores constitute a time-distribution with respect to this event, and it is very likely that it is this element which is principally reflected in response curves.

These findings obtained more quantitative expression in a study by Vary and Halvorson,[15] who also utilized darkening under phase contrast to monitor germination. The germinatinon time of a spore was taken to be that interval from first contact with the germinant to completion

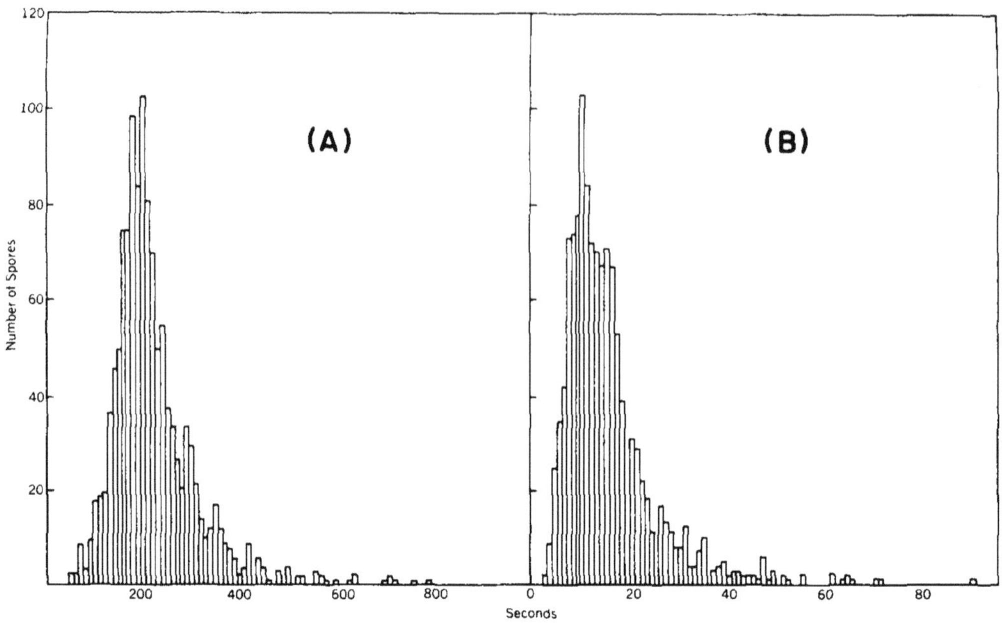

FIGURE 3. Density distributions of microlag (A) and microgermination (B) times on darkening under phase contrast microscopy in spores of *B. cereus* T. (From Vary, J. C. and Halvorson, H. O., *J. Bacteriol.*, 89, 1340, 1965. With permission.)

of the observed change. This time was further resolved into an initial interval of no change, called the microlag time, followed immediately by the interval of change itself, called the microgermination time. Over a large enough sample, both microlag and microgermination times were found to be distributed over a range of values, as can be seen in Figure 3. For the sample shown, the range of microlag times was found to be about ten times that of the microgermination times, reiterating the dominant role played by the asynchrony in the spore sample in determining the shape of a response curve. Statistical analysis of these two distributions failed to establish correlation between them. Independence between microlag and microgermination times in a spore was further suggested by the observation that changes in heat-activation times and germinant concentrations result in changes in the microlag times, but leave the microgermination times unaffected. Both distributions are affected by changes in incubation temperature, however.

C. The Microgermination Function

The use of a microscope photometer to record changes in refractility and increased light transmission in individual spores has revealed that microgermination can occur in a complex way depending on the choice of environmental conditions and spore pretreatment. As observed by Hashimoto et al.[16] in their first study on germination of the single spore, two distinct phases can be differentiated in the microgermination curve. In *B. cereus* T spores, heat-activated at 65°C for various intervals of time and germinated at 25°C in L-alanine (5 mg/mℓ), adenosine (2.5 mg/mℓ), and Tris buffer (0.05 *M*, pH 8.3), there was a relatively rapid initial drop of between 50 and 70% in refractility in an interval of about 75 sec, followed by a more gradual decrease lasting for 3 min or more. Fresh spores, if not heat-activated, had considerably longer microlag times than heat-activated spores. However, once initiated, the first phase of microgermination was completed within a rather narrow time range, regardless of heat-activation or degree of aging. The duration of the second phase of microgermination seemed slightly reduced by heat-activation. Variations in the pH of the medium altered both the microlag times and the second phase of microgermination considerably, but the first phase of microgermination seemed little affected. It was concluded that microgermination is probably a composite of two independent phenomena occurring in sequence, without it being possible to determine whether termination of the first is a prerequisite for initiation of the second.

Variations in incubation temperature were shown to affect microlag time and the extent of its distribution, both decreasing as the temperature increased from 16 to 30°C.[17] Decreases in each phase of microgermination were also observed to accompany temperature increases, agreeing with the observations of Vary and Halvorson. The second phase of microgermination was almost completely blocked at 43°C, the core of the spores remaining semirefractile until the temperature was reduced to 38°C or below, whereupon the spores completed darkening under phase contrast. Also in agreement with Vary and Halvorson,[15] microgermination curves were found to be very similar over a wide range of L-alanine concentrations (0.05 to 50 mg/mℓ), whereas the microlag times were considerably reduced by increasing concentrations of this germinant. The possibility of independence between the two phases of microgermination was reinforced by the observation that Ca^{2+} ions, in concentrations from 0.2 to 0.4 *M* blocked the second phase of microgermination reversibly. In addition, the first phase was lengthened considerably. Thus, the various stages of microgermination of individual spores (i.e., microlag, the first and second phases of microgermination) appear to have different sensitivities to factors affecting germination. From their studies, these authors concluded that the shape of a germination curve is determined largely by the heterogeneity in the microlag times.

As was noted earlier, heat-activation of spores, i.e., their exposure to elevated but sublethal temperatures, generally results in an increased extent and rate of germination, as well as in a simplification of the requirements for initiation of this process. This stimulus appears to affect principally those mechanisms which determine the heterogeneity of spore samples. At sufficiently high, but still sublethal temperatures of heat-activation, however, there would appear to be damage such that both the microlag distribution and the kinetics of microgermination are affected.[18] These effects can be such that the sample response curve is no longer dominated by the microlag distribution, but exhibits features of bimodalism.[19,20]

III. EMPIRICAL FORMULATIONS

Loss of absorbance in a sample of germinating bacterial spores has been widely utilized in monitoring this process because of its convenience as well as the fact that it provides generally reproducible results. Given y_0 and y_∞ as the initial and final absorbances, respectively, of a sample of germinating spores, and y(t) its value at some intermediate time, Hachisuka et al.[21] proposed that the relative absorbance, G(t), given by

$$G(t) = \frac{y(t) - y_\infty}{y_0 - y_\infty} \qquad (1)$$

is appropriate in studies on germination rate, since it is independent of the spore fraction which eventually germinates, assuming the spore concentration is within the range in which it is linearly dependent on the absorbance. Following an initial interval of no change, called the sample lag time, G(t) decreases sigmoidally from unity to zero. Various ways of presenting absorbance data can be found in the literature, but these can easily be reformulated into G(t).

Early interest in the effects of various pretreatments of spores, as well as of modifications in the germination environment, focused on the slope of the germination curve at its point of inflection. Woese et al.[22] proposed that following the initial lag period, the germination curve could be reasonably approximated by an exponential dependence on time about its point of inflection. The argument of this exponential function could then serve to characterize the sample in question, with variations in the experimental conditions being reflected in changes in this parameter. Of course, such a formulation could describe only a limited portion of the experimental curve.

An expression capable of reproducing a much larger portion of the absorbance vs. time curve in *B. cereus* T spores was formulated by McCormick from the observation that a plot of ln ln 1/Y(t) vs. ln t, where Y(t) = 1 − G(t), results in a linear relationship with a negative slope.[23] From

$$\ln \ln\left(\frac{1}{Y(t)}\right) = -c \ln t + \ln \ln\left(\frac{1}{Y_0}\right) \tag{2}$$

where Y_0 = Y at t = 1 unit of time and −c is the slope of the resulting line, one obtains

$$Y(t) = e^{-kt^{-c}} \tag{3}$$

where k = ln $1/Y_0$. This expression generates a sigmoid from zero to unity as t increases from zero to infinity. As was demonstrated in this paper, the parameters k and c, as well as the final sample absorbance, y_∞, can be evaluated from a knowledge of y_0 and the determination of the absorbances at three successive points in time. The slope of the curve at its point of inflection is easily evaluated if k and c are known. Different temperatures of incubation, activation times at 65°C, and changes in the concentration of the germinant (L-alanine) suggested that k and c react differently to changes in germination conditions. Thus, c appeared to have its maximum at 25°C, while k decreased continuously with increases in incubation temperature over the range from about 10 to 40°C. On the other hand, c was little affected by increases in either the time of heat-activation or in the concentration of the germinant, while k decreased in response to both of these changes. These results suggest that k and c may indeed be independent parameters. Interestingly, the average microlag and microgermination times responded to heat-activation and changes in the germinant concentration in a similar manner to k and c, respectively, suggesting a possible association of k with the heterogeneity of the spore suspension, and of c with the average microgermination time. However, c does have a maximum with respect to incubation temperature, while the average microgermination time does not.

IV. MATHEMATICAL MODELS

A. The Woese Hypothesis

The nature of the mechanism by which the spore interacts with the germinant to produce a cell irreversibly committed to initiate germination remains unresolved. Woese et al.[24] have proposed that it might have a biochemical basis. In this model, a spore is supposed to contain n molecules of a germination enzyme which reacts in the presence of a germinant by synthesizing an initiator protein. When the concentration, P, of this protein reaches a critical value, P_c, the spore germinates. The rate of production of this protein is given by:

$$\frac{dP}{dt} = Kn \tag{4}$$

where K is the reaction-rate constant. All spores in the sample containing n enzyme molecules will remain unchanged for a critical time $t_c = P_c/Kn$, at which point they germinate. Assuming that the fraction of spores characterized by each value of n can be determined from a Poisson distribution, a plot of the fraction germinated vs. time yields a curve which increases from zero discontinuously. Application of absorbance vs. time data to the behavior predicted by this model led these investigators to conclude that their spores each contained, on average, nine enzyme molecules. To account for less than complete germination, as observed under some initial conditions, an expanded model was developed in which a degree of lability is conferred on the initiator protein. Thus, the rate equation above was rewritten as

$$\frac{dP}{dt} = Kn - k_2P \tag{5}$$

where k_2 is the rate constant for the breakdown of the initiator protein. In this way the initiator protein concentration builds up to a steady-state value P_s, such that

$$P = \frac{Kn}{k_2}\left(1 - e^{-k_2t}\right) \tag{6}$$

$$P_s = \frac{Kn}{k_2} \tag{7}$$

Those spores for which P_c is greater than P_s will not germinate. The extent of germination is then determined, at least in part, by the dependence of K and k_2 on the environmental conditions. Another way of changing P_s is to alter the effective value of n. Thus, one could visualize heat shock as somehow activating the enzyme molecules. In a similar way, this model can be made to account, at least qualitatively, for many features observed in germination.

The Woese model correctly attributes the role of principal determinant of the shape of a germination curve (e.g., loss of absorbance vs. time) to the time distribution of the spores for initiation of the observed change.[15] In proposing that this asynchrony is generated by events resulting from the interaction of spores and germinant, these authors introduce a second key element into the eventual development of a comprehensive theoretical treatment of this phenomenon. However, the trigger mechanism proposed, as well as the assumption that loss of absorbance is its immediate consequence, cannot accommodate certain features of the germination process.

First, the rate at which the spores of a sample traverse the threshold concentration, P_c, is determined by the values of K and k_2, as well as the manner in which the spores are distributed over n. This rate is thus independent of the measurable property selected for observation in an experiment, so that one should expect the same response curve on all of them. Faced with the fact that the measurable events in germination occur as a time-ordered sequence, and that they may differ in their time evolution as well,[10] it is evident that the asynchrony generated by a trigger mechanism must undergo modification as the sample of spores progresses through the various events which make up germination.

Second, the triggering mechanism advanced in the Woese hypothesis features an intrinsic lag time. No spores germinate until those containing the maximum number of germination enzyme molecules reach the critical concentration in initiator protein, P_c. Recent experiments have shown that commitment, or triggering, commences immediately on contact between spores and germinant.[25] Assuming that the critical time, t_c, for that fraction of spores containing the maximum number of enzyme molecules can be made negligibly small would require, according to Equation 6, that P_c be zero.

B. The Transition-State Model

The entire absorbance vs. time curve of a sample of germinating bacterial spores can be

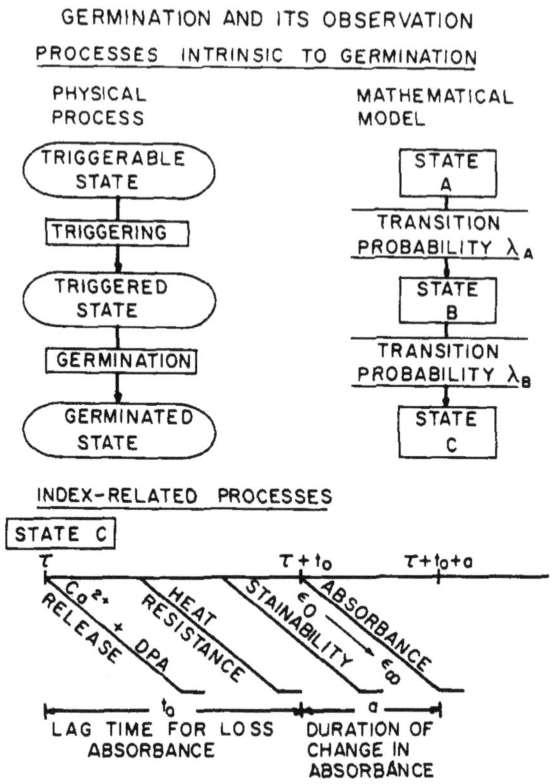

FIGURE 4. Schematic diagram of the germination process and its observation. (From Lefebvre, G. M. and Antippa, A. F., *J. Theor. Biol.*, 95, 489, 1982. With permission.)

reproduced accurately by a model which is also qualitatively consistent with the currently accepted description of this process.[26] As can be seen in its schematic representation shown in Figure 4, the rate of entry of spores into the phase of degradative events is determined by transitions between three spore states, denoted A, B, and C. The transition probabilities λ_A and λ_B are taken to be independent of time, but can otherwise be functions of the environmental variables. In the presence of the germinant, spores initially in state A undergo triggering to state B. These triggered spores are committed eventually to undergo germination to state C. A spore in state C is irreversibly engaged in the sequence of degradative events.

As was pointed out earlier, the various ways in which germination may be followed arise as a time-ordered sequence during the degradative phase, so that the extent of germination at any instant in a sample would appear to be greater for any early event than for a late one. The germination event cannot depend in this way on the property chosen for its observation. It is reasonable then to consider the spore to have germinated when it reaches the earliest event in this sequence, i.e., on entry into state C itself.

1. Kinetics of Germination

In the context of this model, the kinetics of germination is described by dC/dt, the rate of entry of spores into the phase of degradative events. If the transitions between states are taken to be instantaneous and irreversible, the rates of change in the number of spores in each of these states can be expressed by

$$\frac{dA(t)}{dt} = -\lambda_A A(t) \quad (8)$$

$$\frac{dB(t)}{dt} = -\lambda_B B(t) + \lambda_A A(t) \quad (9)$$

and, assuming that the spores preserve their integrity,

$$A(t) + B(t) + C(t) = A_0 \quad (10)$$

where A_0 is the concentration of spores which eventually germinate. This system of simultaneous differential equations can be solved to yield:

$$A(t) = A_0 e^{-\lambda_A t} \quad (11)$$

$$B(t) = \frac{\lambda_A}{\lambda_B - \lambda_A} A_0[e^{-\lambda_A t} - e^{-\lambda_B t}] \quad (12)$$

$$C(t) = A_0 + \frac{A_0}{\lambda_B - \lambda_A} [\lambda_A e^{-\lambda_B t} - \lambda_B e^{-\lambda_A t}] \quad (13)$$

The specific rate of germination is then

$$\frac{1}{A_0}\frac{dC(t)}{dt} = \frac{\lambda_A \lambda_B}{\lambda_B - \lambda_A} [e^{-\lambda_A t} - e^{-\lambda_B t}] \quad (14)$$

2. *Manifestations of Germination*

The above description of germination is independent of which characteristic property is selected to follow the time course of this process. These observable changes in the properties of the spore are consequent to its having entered state C, as illustrated in Figure 4. Let y(t) be the measured value in one such property in a sample of germinating spores a time t after addition of the germinant, and let $\epsilon(t)$ be the contribution to this value by a spore having entered state C at $t = 0$. The time dependence of this coefficient can be represented by the expression

$$\epsilon(t) = \epsilon_0 + (\epsilon_\infty - \epsilon_0) \int_0^t \alpha(t')\, dt' \quad (15)$$

where t' is the variable of integration over the interval from zero to t, and $(\epsilon_\infty - \epsilon_0)\alpha(t')$ describes the time rate of change in ϵ at any time in this interval. The function α is assumed to include the lag time appropriate to the property monitored, and is subject to the normalization condition

$$\int_0^\infty \alpha(t')\, dt' = 1 \quad (16)$$

The spores enter state C asynchronously, their rate of entry at any instant t' in the interval from zero to t being given by $dC(t')/dt'$. The number of spores which enter state C in the interval from τ to $\tau + d\tau$ is just $dC(t')/dt'|_{t'=\tau}d\tau$. Each of these spores contributes $\epsilon(t - \tau)$ to the observed property at time t, so that y(t) can be expressed as

$$y(t) = \epsilon_0(N_0 - A_0) + \epsilon_0[A(t) + B(t)] + \int_0^t \left.\frac{dC(t')}{dt'}\right|_{t'=\tau} \epsilon(t - \tau)\, d\tau \qquad (17)$$

where N_0 is the total spore concentration, $(N_0 - A_0)$ is the fraction of spores which will not germinate in the experiment, and A(t) and B(t) are the concentrations of spores remaining in states A and B at time t, respectively. This last expression can be integrated by parts. A change of variable, $\tau \rightarrow (t - \tau)$ leads to

$$y(t) = y_0 + (\epsilon_\infty - \epsilon_0) \int_0^t C(t - \tau)\, \alpha(\tau)\, d\tau \qquad (18)$$

where $y_0 = \epsilon_0 N_0$, use was made of Equation 10, noting from Equation 13 that $C(0) = 0$ and from Equation 15 that $d\epsilon(t)/dt = (\epsilon_\infty - \epsilon_0)\alpha(t)$. Thus, y(t) is essentially the convolution of C(t) and $\alpha(t)$.

Substituting for C(t) from Equation 13 allows y(t) to be expressed as

$$y(t) = y_0 + A_0(\epsilon_\infty - \epsilon_0)\left[F(t,0) + \frac{\lambda_A}{\lambda_B - \lambda_A} e^{-\lambda_B t} F(t,\lambda_B) - \frac{\lambda_B}{\lambda_B - \lambda_A} e^{-\lambda_A t} F(t,\lambda_A)\right] \qquad (19)$$

where

$$F(t,\lambda) = \int_0^t e^{\lambda\tau}\alpha(\tau)\, d\tau \qquad (20)$$

To evaluate the asymptotic value of y(t), we note that

$$\lim_{t\to\infty} e^{-\lambda t} F(t,\lambda) = 0 \qquad (21)$$

provided that $\alpha(t)$ is zero for times greater than some finite time t_{max}. Moreover, due to the normalization condition on $\alpha(t)$ above,

$$\lim_{t\to\infty} F(t,0) = 1 \qquad (22)$$

Thus,

$$y_\infty = y_0 + A_0(\epsilon_\infty - \epsilon_0) \qquad (23)$$

as expected.

The above integral equation for y(t) can be transformed into a corresponding differential equation. Making use of the fact that

$$\frac{\partial F(t,\lambda)}{\partial t} = e^{\lambda t}\alpha(t) \qquad (24)$$

in the successive differentiation of Equation 19, one obtains

$$F(t,0) = \frac{1}{A_0(\epsilon_\infty - \epsilon_0)\,\lambda_A\lambda_B}\left[\frac{d^2y(t)}{dt^2} + (\lambda_A + \lambda_B)\frac{dy(t)}{dt} + \lambda_A\lambda_B[y(t) - y_0]\right] \quad (25)$$

subject to the boundary conditions

$$y(0) = y_0 = \epsilon_0 N_0$$

$$\left.\frac{dy(t)}{dt}\right|_{t=0} = 0 \quad (26)$$

Combining Equation 25 with Equations 15, 20, and 24, differential expressions for $\epsilon(t)$ and $\alpha(t)$ can be derived:

$$\epsilon(t) = \epsilon_0 + \frac{1}{A_0\lambda_A\lambda_B}\left[\frac{d^2y(t)}{dt^2} + (\lambda_A + \lambda_B)\frac{dy(t)}{dt} + \lambda_A\lambda_B[y(t) - y_0]\right] \quad (27)$$

and

$$\alpha(t) = \frac{1}{A_0(\epsilon_\infty - \epsilon_0)\,\lambda_A\lambda_B}\left[\frac{d^3y(t)}{dt^3} + (\lambda_A + \lambda_B)\frac{d^2y(t)}{dt^2} + \lambda_A\lambda_B\frac{dy(t)}{dt}\right] \quad (28)$$

The above expressions can be used to evaluate $\epsilon(t)$ and $\alpha(t)$ from the measured values of $y(t)$ and its derivatives, provided λ_A and λ_B are known.

3. *Approximations for $\alpha(t)$*

a. Rectangular Pulse

The function $\alpha(t)$ can be determined experimentally by observing the germination of a single spore. In the absence of a complete parametric description, $\alpha(t)$ can be approximated by functions which recognize that $\Delta\epsilon$ has a finite duration and that it occupies a specific position in the sequence of degradative events. The simplest form of the function α meeting these requirements is the rectangular pulse of duration a, as shown in Figure 5. The height of the pulse is fixed at 1/a to satisfy the normalization condition in Equation 16. Thus, identifying all quantities affected by this approximation with the subscript a, we have

$$\alpha_a(t) = \begin{cases} 0 & t < t_0 \text{ and } t > t_0 + a \\ 1/a & t_0 \leq t \leq t_0 + a \end{cases} \quad (29)$$

Substitution into Equation 18 leads to

$$y_a(t) = y_0, \quad t \leq t_0 \quad (30a)$$

$$y_a(t) = y_0 + A_0(\epsilon_\infty - \epsilon_0)\left[\frac{(t - t_0)}{a} + \frac{\lambda_A}{\lambda_B - \lambda_A}\left(\frac{1 - e^{-\lambda_B(t-t_0)}}{a\lambda_B}\right) - \frac{\lambda_B}{\lambda_B - \lambda_A}\left(\frac{1 - e^{-\lambda_A(t-t_0)}}{a\lambda_A}\right)\right], \quad t_0 \leq t \leq t_0 + a \quad (30b)$$

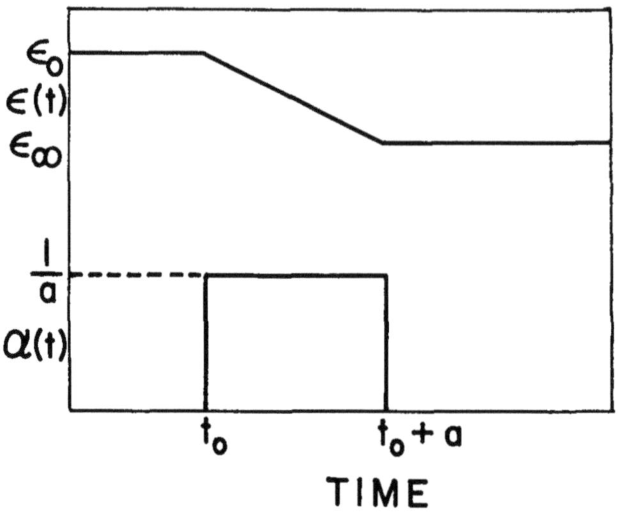

FIGURE 5. A plot of the rectangular pulse approximation for $\alpha(t)$, together with the corresponding function $\epsilon(t)$. (From Lefebvre, G. M. and Antippa, A. F., *J. Theor. Biol.*, 95, 489, 1982. With permission.)

$$y_a(t) = y_\infty + A_0(\epsilon_\infty - \epsilon_0)\left[\frac{\lambda_A}{\lambda_B - \lambda_A}\left(\frac{e^{a\lambda_B} - 1}{a\lambda_B}\right)e^{-\lambda_B(t-t_0)} - \frac{\lambda_B}{\lambda_B - \lambda_A}\left(\frac{e^{a\lambda_A} - 1}{a\lambda_A}\right)e^{-\lambda_A(t-t_0)}\right], \quad t \geqslant t_0 + a \tag{30c}$$

b. Delta Function Approximation

In the limit of zero pulse width, $\alpha_a(t)$ reduces to a Dirac delta function at $t = t_0$, and the change $\Delta\epsilon = \epsilon_\infty - \epsilon_0$ takes place instantaneously. Identifying all quantities in this approximation by the index s, we have

$$\alpha_s(t) = \lim_{a \to 0} \alpha_a(t) = \delta(t - t_0) \tag{31}$$

Substitution of this expression into Equation 18, gives

$$y_s(t) = y_0 + (\epsilon_\infty - \epsilon_0)\, C(t - t_0) \tag{32}$$

Written out explicitly, this becomes

$$y_s(t) = \begin{cases} y_0, & t < t_0 \\ y_\infty + \dfrac{A_0(\epsilon_\infty - \epsilon_0)}{\lambda_B - \lambda_A}\left[\lambda_A e^{-\lambda_B(t-t_0)} - \lambda_B e^{-\lambda_A(t-t_0)}\right], & t \geqslant t_0 \end{cases} \tag{33}$$

Part of the data accumulated from the germination of spores heat-activated at 70°C for 3 min is shown in Figure 6. These data were analyzed for best fit of Equations 30 and 33.

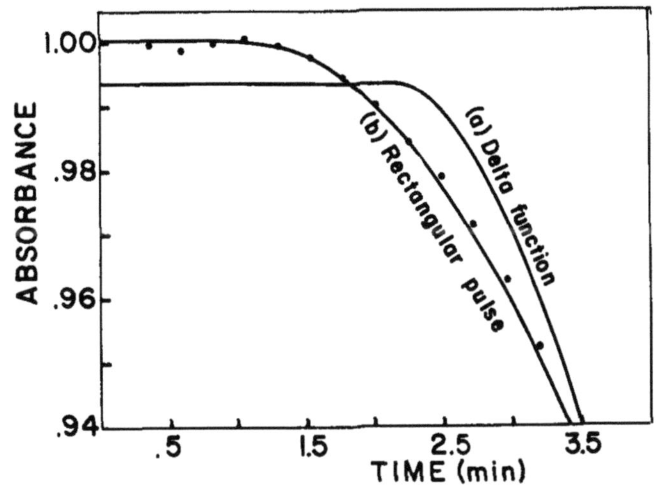

FIGURE 6. An enlarged view of the shoulder region of the absorbance vs. time curve of a sample of germinating *B. megaterium* spores. The spores, heat activated at 70°C for 3 min, were exposed to 10 m*M* β-D-glucose in 50 m*M* phosphate buffer, pH 7.0. The experimental points are plotted as dots. Curve (a) is a fit of Equation 33 corresponding to a delta function approximation for α(t). Curve (b) is a fit of Equations 30 corresponding to a rectangular pulse approximation for α(t). (From Lefebvre, G. M. and Antippa, A. F., *J. Theor. Biol.*, 95, 489, 1982. With permission.)

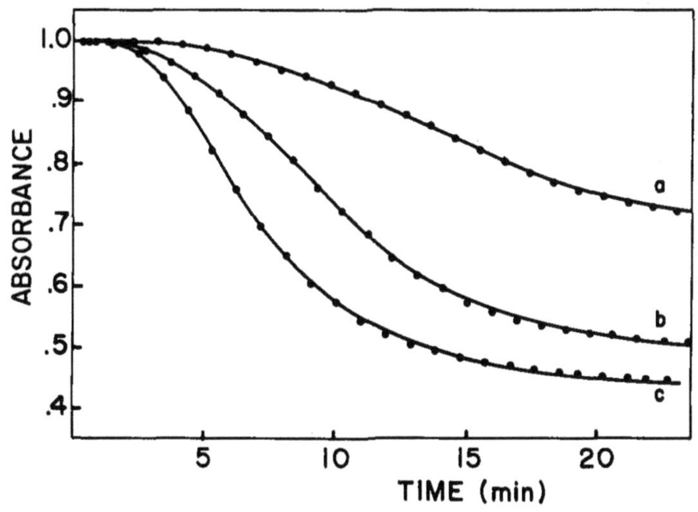

FIGURE 7. Absorbance vs. time curves predicted using Equation 30, corresponding to a rectangular pulse for α(t). Experimental data are plotted as points. The spores were (a) unactivated, (b) heat activated at 60°C for 3 min, and (c) heat activated at 70°C for 3 min. All were germinated under the conditions given in Figure 6. (From Lefebvre, G. M. and Antippa, A. F., *J. Theor. Biol.*, 95, 489, 1982. With permission.)

The predicted curves are shown as solid traces. The fit with the rectangular pulse is clearly superior in this initial portion, while both curves are essentially coincident with the data at all later points. This suggests that this model as a whole provides an accurate description of the change in absorbance due to spore germination. The ability of Equation 30 to reproduce experimental results is illustrated in Figures 7 and 8.

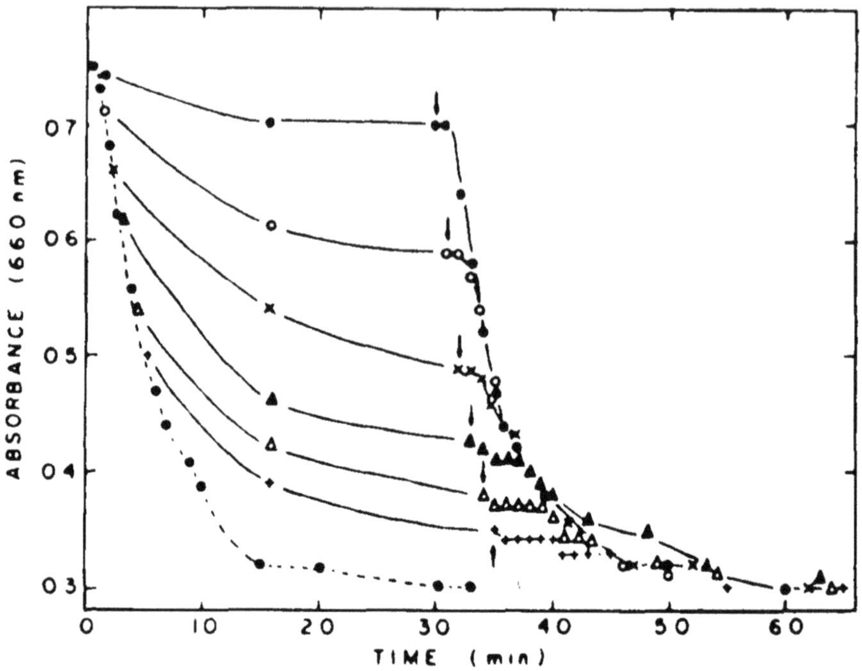

FIGURE 8. Temperature shift-up during initiation of strain JV-10 of *B. megaterium* QM B1551 spores. Heat activated spores (10 min at 60°C) were diluted into 1.5 mℓ of 0.2 *M* glucose, 5 m*M* Tris (pH 8) solution, and incubated at 30°C for 0 (●), 1 (○), 2 (×), 3 (▲), 4 (△), or 5 (+) min. After these preincubations, the temperature was raised rapidly to 46°C by the addition of 1.5 mℓ of the glucose in Tris solution pre-equilibrated to 60°C. The absorbances at 660 nm were measured for 30 min and the tubes were then placed in a 30°C water bath (at the time indicated by the arrows) and the absorbances measured for another 30 min. For the control (●), heat activated spores were held continuously at 30°C. (From Vary, J. C., *J. Bacteriol.*, 121, 197, 1975. With permission.)

V. TRIGGERING

A number of investigations provide evidence consistent with the postulate that the asynchrony in a spore sample for entry into the phase of degradative events is generated by a sequence of transitions between spore states. These deal either implicitly or explicitly with the triggering event. Harrell and Halvorson[6] were the first to isolate this event experimentally. Using environmental conditions which would normally lead to 100% spore germination, these investigators found that acidification of a sample undergoing physiological germination to pH 2.0 completely arrested loss of absorbance. In one experiment, acidification was affected 45 sec after mixing spores and germinant in each of three samples. Two samples were rinsed repeatedly at this low pH to remove the germinant, while the third was maintained at pH 2.0. The initial conditions were fully restored in the first sample, and it proceeded to complete absorbance loss in about 20 min of further incubation. The second sample was also restored in all but one respect: the germinant was omitted. It also required the full 20 min to complete the observed change, but only achieved about 40% of the extent of germination of the first sample. Repeated washings of this second sample did not alter this 40% level of germination. For the sample maintained at pH 2.0, there was no change in absorbance beyond its value at acidification.

All the spores which germinated in the second sample after its neutralization must have been in a state different from the initial spore state since they could then complete this process in the absence of the germinant. The sample lag time on loss of absorbance is no more than a few minutes, so that all spores having reached the initiation phase at the time of acidification would complete absorbance loss very quickly on restoration of neutral pH. Since the spores of the second sample required 20 min to complete absorbance loss after neutralization, it is reasonable to suppose that most of these spores were in a state intermediate between states A and C. Moreover, the spores in this state could not exit from it simultaneously, but had to do so asynchronously. The simplest assumption in this case is that this transition occurs randomly in time.

The degree of acidification utilized by Harrell and Halvorson apparently completely arrested the normal progress of the germinating spores at all stages. In temperature-sensitive mutants, however, the germination process may be arrested in a more localized fashion when the temperature is shifted beyond permissive values. This can be seen in a study of mutants of *B. megaterium* QM B1551,[27] the spores of which were found to germinate normally in the presence of glucose at 30°C, but could not do so at 46°C. In one series of experiments, the spores were first exposed to glucose at 30°C for from 1 to 5 min, immediately followed by rapid shift up to 46°C. As shown in Figure 8, the absorbance in these samples continued to decrease for a time after shift-up to the higher temperature, gradually approaching some final value. Those spores contributing to the absorbance loss after shift-up to 46°C must have been in a state different from the initial spore state, since spores of this mutant exposed to glucose at this high temperature do not germinate. This triggered state cannot be state C, since all spores in this state at the time of the temperature jump require no more than the lag time on absorbance (of the order of 1 or 2 min) to initiate this change, whereas the observed postshift-up decrease in absorbance lasted about 30 min. These results could be accounted for in the transition state model by considering that the temperature block occurs in the intermediate state B.

The rate of commitment to germinate, or triggering, heat-activated *B. megaterium* KM was found to increase (from negative values) exponentially from the time of addition of the germinant L-alanine, as can be seen in Figure 9.[28] The absence of a lag time in this event is consistent with the transition state model. That this leads to a sigmoidal release of Ca^{2+}, the earliest indicative response observed in this strain, would imply the presence of a subsequent random transition. While it is true that Ca^{2+} release in the individual spore is probably not instantaneous, i.e., it has a finite microgermination time, this could not be greater than the sample lag time on absorbance loss, which is of the order of a few minutes. Ca^{2+} release in the sample extends over an interval an order of magnitude larger than this. As with most other measurable changes, the kinetics of Ca^{2+} release in a sample of germinating spores principally reflects its microlag distribution.

Very pursuasive evidence for the existence of an intermediate state in the transition state model can be found in a series of studies on germination of unactivated *B. cereus* T spores.[29,30] While heat-activated spores of this strain will germinate in the presence of either L-alanine or inosine, only a combination of these two germinants is effective without heat-activation.[31] Utilizing a mixture of L-alanine and inosine capable of inducing near 100% germination in unactivated spores, Shibata et al.[29] observed that preincubation of the spores with either of these two substances alone produced opposing results when they were subsequently placed in contact with the effective mixture. Preincubation for various lengths of time with L-alanine alone decreased the time at which the point of inflection occurred in the absorbance vs. time curves obtained on adding inosine to the reaction mixture. The extent of this reduction increased with the time of preincubation in L-alanine. On the other hand, preincubation with inosine alone produced an inhibitory effect. It is significant that both activation by L-alanine and inhibition by inosine do not affect all spores instantaneously, with activation maximal by 45 min, and inhibition extending up to 60 min in these experiments.

In a subsequent investigation, Shibata et al.[30] found that uptake of radioactive L-alanine was maximal in the first minutes after mixing with the spores under conditions identical to

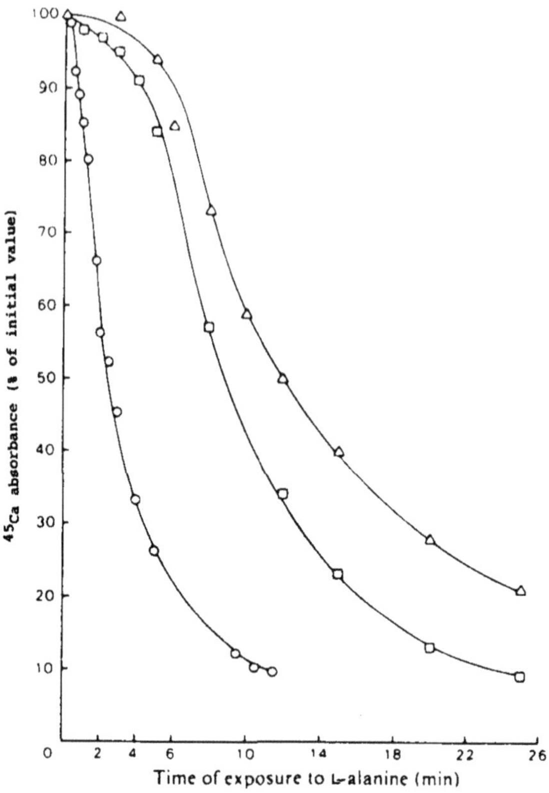

FIGURE 9. Comparison of the rate of absorbance decrease, △, calcium release, □, and commitment, ○, during the germination of *B. megaterium* KM with L-alanine. (From Stewart, G.S.A.B., Johnston, K., Hagelberg, E., and Ellar, D. J., *Biochem. J.*, 198, 101, 1981. With permission.)

those utilized in the preceding study. This can be seen in Figure 10. These spores are not irreversibly engaged in the germinative process, however. Thorough washing to remove L-alanine and subsequent resuspension in the same buffer requires the presence of inosine for subsequent loss of absorbance. Moreover, as shown in Figure 11, the number of spores which eventually germinate depends on the duration of exposure to L-alanine over a range of intervals much greater than that required for uptake saturation. In the context of the transition state model, spores in state A are those which, having taken up L-alanine, are competent to effect an initial irreversible transition from this state in the presence of excess L-alanine. The dependence of the rate and extent of germination on the concentration of the germinant might be explicable in part in terms of its effect on this transition probability. The fact that there are spores which, after removal of L-alanine, are competent to react with inosine to effect transition to the phase of degradative events (state C), testifies very strongly in favor of an intermediate or triggered state. In view of the fact that these spores will germinate in the presence of either L-alanine or inosine when heat-activated makes this an excellent system for the study of the effects of spore pretreatment, including aging.

VI. A GENERALIZED FORMULATION

The shape of the response curve of most observable changes in germination mainly reflects the time distribution of initiation of the relevant change by the individual spores. This is

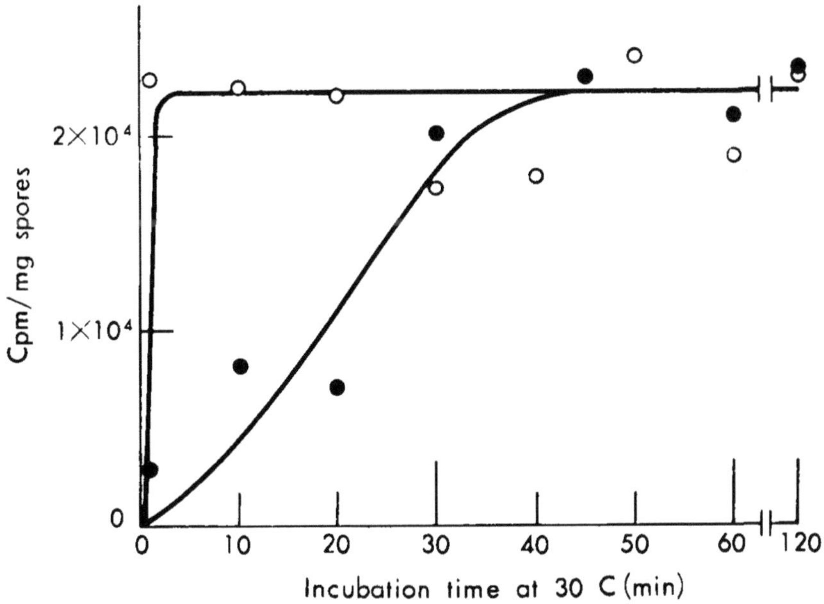

FIGURE 10. Incorporation of L-[^{14}C] alanine into unactivated spores of *B. cereus* T during preincubation. Unactivated spores were incubated at 30°C in 0.1 *M* phosphate buffers (pH 8.0) containing L-[^{14}C] alanine (0.5 μCi/mℓ): ○, sodium phosphate buffer; ●, potassium phosphate buffer. (From Shibata, H., Takamatsu, H., Minami, M., and Tani, I., *Microbiol. Immunol.*, 22, 443, 1978. With permission.)

due to the large difference in the ranges of the microlag and microgermination distributions.[15,16] The transition-state model expressed by Equation 18 provides an accurate representation of this observed time development. It must be recognized, however, that microlag distributions on events sufficiently far removed from one another in time may differ appreciably due to the cumulative effects of the microgermination distributions of intervening events, on the one hand, or to the occurrence of randomizing events, on the other. In general, then, whatever the property whose change is to be observed in germination, the response curve will be influenced to some degree by the relevant microlag and microgermination distributions, by the kinetics of change in the individual spore, and by the position of the observed change in the sequence of events. These constitutive elements can be included in a formalistic expression for the average contribution of a given spore to the observed event at any time t after initiation of the process of germination.

The coefficient $\epsilon(t)$ was introduced earlier as the individual spore's contribution to the property selected for observation. Making use of the simple form of $\epsilon(t)$ in Figure 5 for illustrative purposes, it can be visualized in terms of the microlag and microgermination times (see Figure 12). Considering that there is no correlation between microlag and microgermination distributions in any one event, all other spores of the sample can be assumed to develop in a similar way, independently selecting their microlag and microgermination times from the observed distributions. The contribution to the chosen event at any time t of a randomly selected spore can be represented by

$$Y_t = \epsilon_{T_1}(t - T_0) \qquad (34)$$

where T_0 and T_1 are non-negative and independent random variables. Figure 13 illustrates how particular values τ_0 and τ_1 of T_0 and T_1, respectively, can serve to specify the time-evolution of a given spore. The time t_0 accounts for the initial interval of no change in the

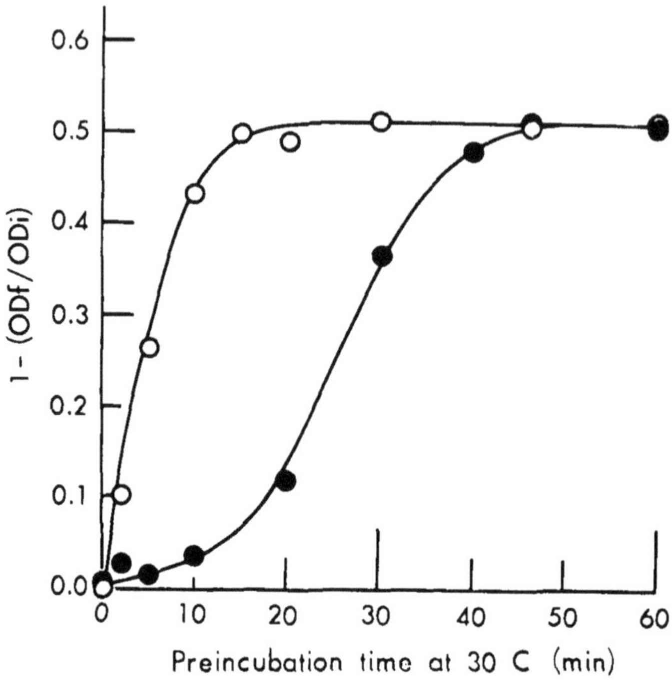

FIGURE 11. Effect of preincubation time with L-alanine on the extent of subsequent germination of spores of *B. cereus* T. Unactivated spores were incubated at 30°C for the times indicated in 0.1 *M* phosphate buffer (pH 8.0) containing 0.5 m*M* L-alanine (○, sodium phosphate buffer; ●, potassium phosphate buffer). After centrifugation the spores were transferred to 0.1 *M* sodium phosphate buffer (pH 8.0) containing 0.1 m*M* inosine. Germination was monitored for 60 min at 30°C, with OD_I and OD_F the initial and final absorbances in an experiment, respectively. (From Shibata, H. J., Takamatsu, H., Minami, M., and Tani, I., *Microbiol. Immunol.*, 22, 443, 1978. With permission.)

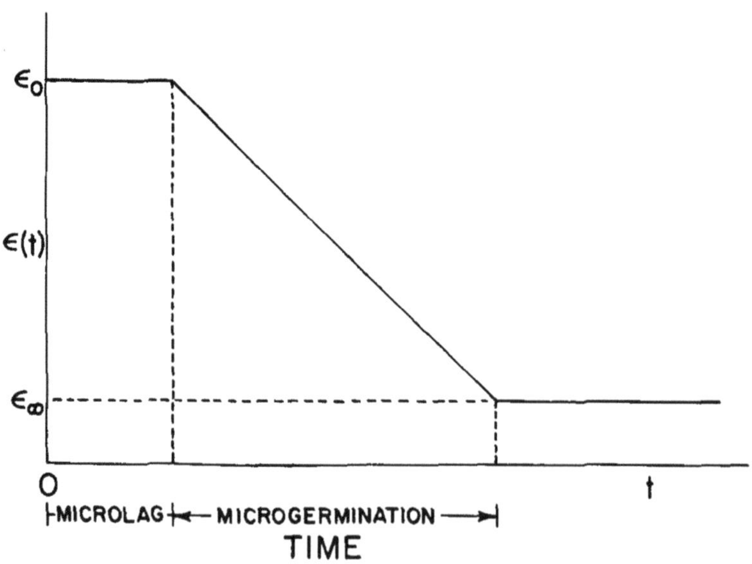

FIGURE 12. The germination time of an individual spore is considered as an initial interval of no change (microlag time) followed immediately by the period of change in the observed index of germination (microgermination time). (From Leblanc, R. and Lefebvre, G.M., *Bull. Math. Biol.*, 46, 447, 1984. With permission.)

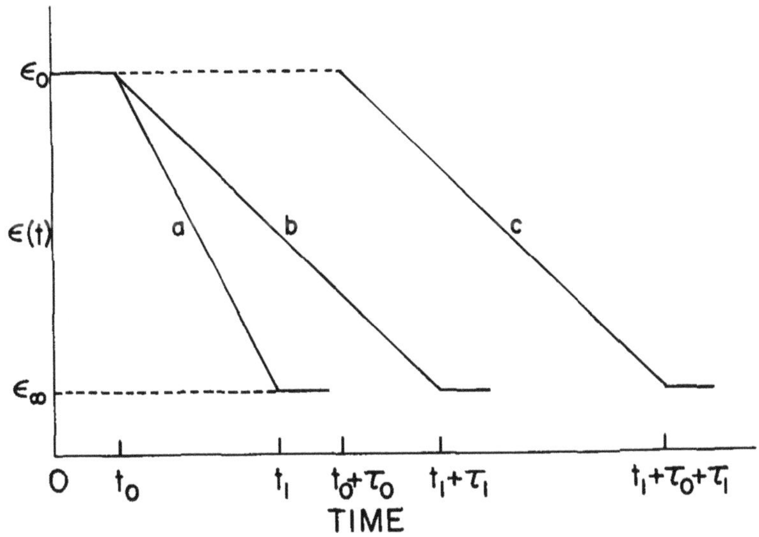

FIGURE 13. The random variables T_0 and T_1 serve to differentiate between individual spore germination times. Spore (a) has the fastest germination time ($\tau_0 = \tau_1 = 0$). Spore (b) has the same microlag as spore (a) but a longer microgermination time. Spore (c) has a longer microlag than either (a) or (b), but the same microgermination time as (b). (From Leblanc, R. and Lefebvre, G. M., *Bull. Math. Biol.*, 46, 447, 1984. With permission.)

property selected for observation, and can be considered as the microlag time of the earliest spore to initiate the observed change. The effect of increasing τ_0 is to shift the curve to later times, while an increase in τ_1 results in a more gradual change.

Because T_0 and T_1 are independent, the spores of a sample can be classified on the basis of microgermination time, the fraction of spores in each class determined by $f_1(\tau_1)$, the density function of the random variable T_1. Each of these classes can be further described by the same density function $f_0(\tau_0)$ of the random variable T_0. For a given value of τ_1, a superposition of individual spore germination curves could be represented as in Figure 14. A similar figure could be drawn for a class having a different value of τ_1 which, in this schematic representation, would differ in the slope of the lines between ϵ_0 and ϵ_∞. As a first approximation in the present development, the limits of ϵ_0 and ϵ_∞ are taken to be the same for all spores of the sample.

As has been shown in detail elsewhere,[32] an expression for the average contribution of a spore to the observed index can be formulated by first considering the probability of possible values Y_t of this property for a selected microgermination time. These probabilities are expressed in terms of $f_0(\tau_0)$ and its cumulative function F_0, defined by

$$F_0(u) = \int_{-\infty}^{u} f_0(\tau_0)\, d\tau_0 \qquad (35)$$

From these can be formulated a conditional distribution function $G_t(y|T_1 = \tau_1)$ which, when combined with the density function $f_1(\tau_1)$, leads to the mean or expected value of Y_t:

$$E(Y_t) = \epsilon_0 + \int_0^{\infty} f_1(\tau_1) \int_{t_0}^{t_1+\tau_1} F_0(t - w)\, \epsilon'_{\tau_1}(w)\, dw\, d\tau_1 \qquad (36)$$

where $\epsilon'_{\tau_1}(w)$ is the first derivative of $\epsilon_{\tau_1}(w)$ with respect to w.

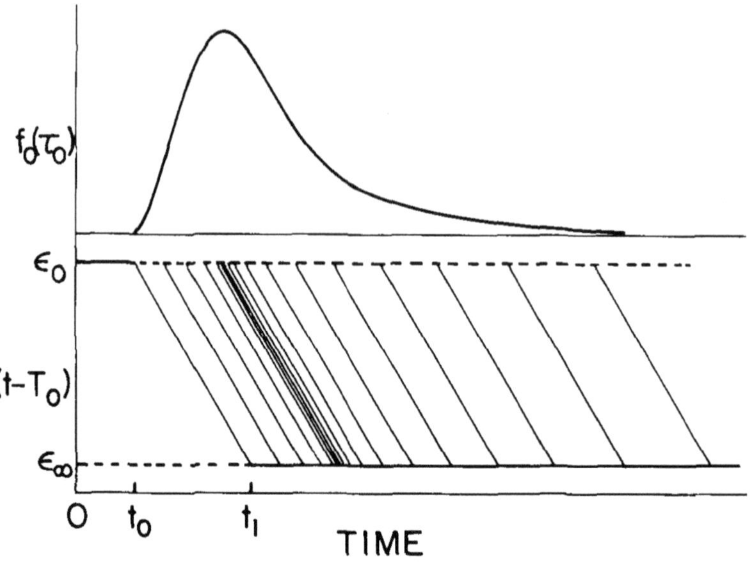

FIGURE 14. Superposition of the schematized germination curves of those spores of the sample with the same microgermination time, i.e., the same τ_1. The correspondence between the density distribution function $f_0(\tau_0)$ of microlag times and the linear density of points on the time axis at ϵ_0 is also illustrated. (From Leblanc, R. and Lefebvre, G. M., *Bull. Math. Biol.*, 4, 447, 1984. With permission.)

Only those spores that initiate the change being monitored contribute to $E(Y_t)$. If the concentration of these is A_0 in a sample of concentration N_0, the value of the corresponding sample measurement at time t after addition of the germinant can be expressed as

$$y(t) = (N_0 - A_0)\,\epsilon_0 + A_0 E(Y_t) \tag{37}$$

where A_0, ϵ_0, and $E(Y_t)$ are specific to the property being measured. Since $E(Y_t)$ is ϵ_0 for $t \leq t_0$ and ϵ_∞ for $t = \infty$, Equation 1 becomes

$$G(t) = \frac{E(Y_t) - \epsilon_\infty}{\epsilon_0 - \epsilon_\infty} \tag{38}$$

where we have utilized $y_0 = \epsilon_0 N_0$ and $y_\infty = y_0 - A_0(\epsilon_0 - \epsilon_\infty)$ as the initial and final observed values, respectively. Regardless of which of ϵ_0 and ϵ_∞ is the larger, G(t) decreases with time from unity for $t \leq t_0$ to zero as t approaches infinity, and is independent of A_0/N_0.

Experimentally derived response curves generally have sigmoid shapes. The various models which might be proposed for one or another of the constitutive elements in $E(Y_t)$, then, need to be consistent with the boundary conditions it must satisfy. Various families of cumulative functions can satisfy Equation 35 for $F_0(\tau_0)$, as well as a similar expression for $F_1(\tau_1)$. Considering only smoothly varying functions, three distinct families can be recognized. These range from curves which increase sigmoidally from zero to unity either with zero or positive initial slope, to curves whose slopes continuously decrease from some positive initial value. In the analysis which follows, we utilize $\epsilon_0 > \epsilon_\infty$ though the results are equally applicable for $\epsilon_0 < \epsilon_\infty$.

Integration of Equation 36 by parts leads to

$$E(Y_t) = \epsilon_0 + \int_0^\infty f_1(\tau_1)\left[F_0(t - w)\,\epsilon_{\tau_1}(w)\Big|_{t_0}^{t_1+\tau_1} + \int_{t_0}^{t_1+\tau_1} f_0(t - w)\,\epsilon_{\tau_1}(w)\right] d\tau_1 \tag{39}$$

or

$$E(Y_t) = \epsilon_0[1 - F_0(t - t_0)] + \epsilon_\infty F_s(t - t_l) + K(t) \quad (40)$$

where

$$K(t) = \int_0^\infty f_l(\tau_l) \int_{t_0}^{t_l+\tau_l} f_0(t - w)\, \epsilon_{\tau_l}(w)\, dw\, d\tau_l \quad (41)$$

and

$$F_s(t - t_l) = \int_0^\infty f_l(\tau_l)\, F_0(t - t_l - \tau_l)\, d\tau_l \quad (42)$$

can be shown to be a cumulative function. Since $\epsilon_{\tau_l}(w)$ is bounded, and since $f_0(t - w)$ is positive or zero over the range of w, it is possible to formulate the inequality

$$\epsilon_\infty M(t) \leq K(t) \leq \epsilon_0 M(t) \quad (43)$$

where

$$M(t) = \int_0^\infty f_l(\tau_l) \int_{t_0}^{t_l+\tau_l} f_0(t - w)\, dw\, d\tau_l \quad (44a)$$

or

$$M(t) = F_0(t - t_0) - F_s(t - t_l) \quad (44b)$$

At sufficiently large values of t, both F_0 and F_s approach unity, while each approaches zero as t approaches t_0. Thus,

$$\lim_{t \to t_0} E(Y_t) = \epsilon_0 \quad (45)$$

and

$$\lim_{t \to \infty} E(Y_t) = \epsilon_\infty \quad (46)$$

The expected end points are recovered on the basis that F_0 and F_l are cumulative functions and that $\tau_l(w)$ is bounded.

Taking the first derivative of $E(Y_t)$ with respect to time, one obtains

$$E'(Y_t) = \int_0^\infty f_l(\tau_l) \int_{t_0}^{t_l+\tau_l} f_0(t - w)\, \epsilon'_{\tau_l}(w)\, dw\, d\tau_l \quad (47)$$

Since a nonmonotonic change in ϵ_{τ_l} can be viewed as a superposition of oppositely directed monotonic processes, we take $\epsilon'_{\tau_l}(w) \leq 0$ for all values of w, given the presumption that

$\epsilon_0 > \epsilon_\infty$. Where $\Delta\epsilon$ is of finite duration, $\epsilon'_{\tau_1}(w)$ can be taken as bounded:

$$a \leqslant \epsilon'_{\tau_1}(w) \leqslant b \tag{48}$$

The case of $\Delta\epsilon$ instantaneous will be treated separately. Thus, it is possible to write the inequality

$$a\ M(t) \leqslant E'(Y_t) \leqslant b\ M(t) \tag{49}$$

where M(t) is as expressed in Equations 44. Since M(t) has already been seen to approach zero in the limit at $t \to t_0$ and $t \to t_\infty$, it is evident that

$$\lim_{t \to t_0} E'(Y_t) = 0 \tag{50}$$

and

$$\lim_{t \to \infty} E'(Y_t) = 0 \tag{51}$$

In the case of finite microgermination time, zero initial and final slopes in the sample response curves are to be expected whatever the detailed nature of the distribution function F_0. Such is not the case, however, where microgermination is instantaneous. In this case Equations 36 and 47 can easily be shown to become

$$E(Y_t) = \epsilon_0 - (\epsilon_0 - \epsilon_\infty)\ F_0(t - t_0) \tag{52}$$

and

$$E'(Y_t) = -(\epsilon_0 - \epsilon_\infty)\ f_0(t - t_0) \tag{53}$$

respectively. A particular example of instantaneous change in ϵ is loss of heat resistance. Considering that a spore is either sensitive to exposure to elevated temperatures or it is not, $\epsilon_0 = 0$ and $\epsilon_\infty = 1$. Equation 38 then becomes

$$G(t) = 1 - F_0(t - t_0) \tag{54}$$

and F_0 completely determines the shape of the sample response curve of this property. Experimentally derived curves of loss of heat resistance are always sigmoidal and generally have a zero initial slope, indicating that this is the shape of the microlag distribution at that point. It is also reasonable to assume that this is the general form of the time distribution of spores in the sample of later events, possibly altered by the microgermination distributions of successive events.

A sigmoidal response curve can be expected to have zero initial and final slopes. Since the G(t) are asymptotic to zero as t becomes very large, the second derivative of $E(Y_t)$ should also be zero there. That $E''(Y_t)$ must be zero at $t = t_0$ as well can be appreciated by considering that the slope in the response curve is zero in the interval $0 \leqslant t \leqslant t_0$, i.e., during the sample lag period. It is now possible to show that $F_0(\tau_0)$ on these indices must be sigmoidal with zero initial slope. Taking the derivative of $E'(Y_t)$ with respect to time, one obtains

$$E''(Y_t) = \int_0^{\infty} f_1(\tau_1) \int_{t_0}^{t_1+\tau_1} f_0(t - w)\, \delta(t - w)\, \epsilon'_{\tau_1}(w)\, dw\, d\tau_1$$

$$+ \int_0^{\infty} f_1(\tau_1) \int_{t_0}^{t_1+\tau_1} f'_0(t - w)\, \epsilon'_{\tau_1}(w)\, dw\, d\tau_1 \qquad (55)$$

where the first integral in this expression takes into account the possible discontinuity in $f_0(t - w)$. This expression can be rewritten as

$$E''(Y_t) = f_0(0) \int_{t-t_1}^{\infty} f_1(\tau_1)\, \epsilon'_{\tau_1}(t)\, d\tau_1 + L(t) \qquad (56)$$

where L(t) is the second integral on the right in Equation 55 expressing integration over the continuous portion of $f_0'(t - w)$ and $f_0(0)$ is the magnitude of the discontinuity in F_0 at $t = t_0$. The lower limit in the integral over τ_1 in Equation 56 has been adjusted from zero in recognition of the fact that smaller values of τ_1 are excluded by the integration over w in the preceding equation. Recalling that $\epsilon'_{\tau_1}(w) \leq 0$ in our working example, and considering that for sufficiently large values of t that $f_0'(t - w)$ will be either negative or zero for all possible values of w, it is possible to formulate the inequality

$$b\, N(t) \leq L(t) \leq a\, N(t) \qquad (57)$$

where

$$N(t) = \int_0^{\infty} f_1(\tau_1) \int_{t_0}^{t_1+\tau_1} f'_0(t - w)\, dw\, d\tau_1 \qquad (58)$$

However,

$$N(t) = f_0(t - w) - \int_0^{\infty} f_1(\tau_1)\, f_0(t - t_1 - \tau_1)\, d\tau_1 \qquad (59)$$

Since both of these density functions approach zero for sufficiently large values of t,

$$\lim_{t\to\infty} L(t) = 0 \qquad (60)$$

The remaining term in Equation 56 also becomes zero in this limit, so that

$$\lim_{t\to\infty} E''(Y_t) = 0 \qquad (61)$$

In the region $t_0 < t < t_1$, Equation 55 becomes

$$E''(Y_t) = f_0(0) \int_0^{\infty} f_1(\tau_1)\, \epsilon'_{\tau_1}(t)\, d\tau_1 + P(t) \qquad (62)$$

where

$$P(t) = \int_0^{\infty} f_1(\tau_1) \int_{t_0}^{t} f'_0(t - w)\, \epsilon'_{\tau_1}(w)\, dw\, d\tau_1 \qquad (63)$$

Invoking the boundedness of $\epsilon'_{\tau_1}(w)$ and considering values of t near enough to t_0 over which $f_0'(t - w)$ does not change sign, it is possible to write

$$a\,R(t) \leqslant P(t) \leqslant b\,R(t) \tag{64}$$

where

$$R(t) = \int_0^\infty f_1(\tau_1) \int_{t_0}^{t} f_0'(t - w)\, dw\, d\tau_1 \tag{65}$$

However,

$$R(t) = f(t - t_0) - f_0(0) \tag{66}$$

which, in the limit $t \rightarrow t_0$, is a zero sum. Therefore,

$$\lim_{t \to t_0} E''(Y_t) = f_0(0)\, \overline{\epsilon'_{\tau_1}(t_0)} \tag{67}$$

As was pointed out earlier, $E''(Y_t)$ should be zero at $t = t_0$ for sigmoidal response curves. From Equation 67, this implies that either or both of $f_0(0)$ and $\overline{\epsilon'_{\tau_1}(t_0)}$ must be zero for those events for which $\Delta\epsilon$ is of finite duration. Where $\Delta\epsilon$ is instantaneous, a zero slope at $t = t_0$ in the sample response curve implies that $f_0(0)$ is zero, as can be seen from Equation 53. The manner in which the time distribution of spores in a sample evolves during the process of their transformation from dormant to vegetative form is the important indicator of the occurrence of significant events in this evolution. The preceding analysis suggests the types of measurements which should be considered in such studies.

If the response curves of the properties commonly utilized to monitor germination are sigmoidal, triggering has been shown to follow a time course with no sample lag time and a nonzero initial slope. In the context of the present formulation, this implies that the change in the corresponding coefficient ϵ is instantaneous, and that G(t) is completely determined by the $F_0(t)$ appropriate to this event. At sufficiently dilute germinant concentrations, triggering in unactivated bacterial spores follows a sigmoidal time course with nonzero initial slope. The point of inflection disappears on increasing the germinant concentration.[25] These results are consistent with the identification of triggering as the earliest of the transitions in the transition state model. The study of the dependence of this transition probability on the nature and concentration of germinants, as well as of the effects of various forms of activation should shed considerable light on this very important mechanism.

VII. CONCLUDING REMARKS

The transition state model proposes that the time distribution of bacterial spores on germination is determined by a sequence of transitions between spore states, the final one of which is identified with the entry of the spore into the phase of degradative events from which arise the various indices by which this process can be monitored. As the spores evolve through this phase, their time distribution can be expected to undergo modification as a result of the rather drastic changes which take place. In principle, knowledge of the microlag distribution of various measurable events in the sequence could serve as a basis for formulating a description of the evolution of the spore through this phase.

To appreciate the contribution of the microgermination distribution of one event to the microlag distribution of some later event, consider that if $F_0(\tau_0)$ and $F_1(\tau_1)$ are appropriate to this earlier index, then

$$F_s(t - t_1) = \int_0^\infty f_1(\tau_1)\, F_0(t - t_1 - \tau_1)\, d\tau_1 \tag{68}$$

is the distribution function for its completion. In a sequence of dependent events, $F_s(t - t_1)$ would also be the microlag distribution for initiation of the next event, and could differ from $F_0(t)$ in two ways. First, the form of the distribution $F_0(t)$ would be altered to a degree dependent on the relative ranges of the microlag and microgermination distributions. While alteration in spore asynchrony may be negligible for short sequences of events, it can become significant, given sufficient separation between observable changes. Second, $F_0(t - t_0)$ describes the time distribution of the spores on initiation of the observed change. As a cumulative function, it is zero for negative values of its argument, i.e., for $t < t_0$. Since the corresponding $F_s(t - t_1)$ describes initiation on the next event in the dependent sequence, then its sample lag time is t_1, the earliest time of completion of the preceding change. Thus, the difference in sample lag times between these two consecutive events is $t_1 - t_0$, the minimum microgermination time of the earlier event. The sample lag time on any event can now be viewed as the summation over all preceding events of their minimum microgermination times.

REFERENCES

1. **Halvorson, H. O., Vary, J. C., and Steinberg, S.,** Developmental changes during the formation and breaking of the dormant state in bacteria, *Ann. Rev. Microbiol.*, 20, 169, 1966.
2. **Keynan, A.,** Spore structure and its relations to resistance, dormancy, and germination, in *Spores VII*, Chambliss, G. and Vary, J. C., Eds., American Society for Microbiology, Washington, D.C., 1978, 43.
3. **Setlow, P.,** Biochemistry of bacterial forespore development and spore germination, in *Sporulation and Germination*, Levinson, H. S., Sonenshein, A. L., and Tipper, D. J., Eds., American Society for Microbiology, Washington, D.C., 1981, 13.
4. **Vary, J. C.,** Bacterial spore germination, *Spore Newslett.*, 6, 191, 1979.
5. **Keynan, A. and Evenchik, Z.,** Activation, in *The Bacterial Spore*, Gould, G. W. and Hurst, A., Eds., Academic Press, New York, 1969, chap. 10.
6. **Harrell, W. K. and Halvorson, H.,** Studies on the role of L-alanine in the germination of spores of *Bacillus terminalis*, *J. Bacteriol.*, 69, 275, 1955.
7. **Halmann, M. and Keynan, A.,** Stages in germination of *Bacillus licheniformis*, *J. Bacteriol.*, 84, 1187, 1962.
8. **Racine, F. M., Dills, S. S., and Vary, J. C.,** Glucose-triggered germination of *Bacillus megaterium* spores, *J. Bacteriol.*, 138, 442, 1979.
9. **Scott, I. R., Stewart, G. S. A. B., Koncewicz, M. A., Ellar, D. J., and Crafts-Lighty, A.,** Sequence of biochemical events during germination of *Bacillus megaterium* spores, in *Spores VII*, Chambliss, G. and Vary, J. C., Eds., American Society for Microbiology, Washington, D.C., 1978, 95.
10. **Levinson, H. S. and Hyatt, M.,** Sequence of events during *Bacillus megaterium* spore germination, *J. Bacteriol.*, 91, 1811, 1966.
11. **Uehara, M. and Frank, H. A.,** Sequence of events during germination of Putrefactive Anaerobe 3679 spores, *J. Bacteriol.*, 94, 506, 1967.
12. **Hsieh, L. K. and Vary, J. C.,** Peptidoglycan hydrolysis during initiation of spore germination in *Bacillus megaterium*, in *Spores VI*, Gerhardt, P., Costilow, R. N., and Sadoff, H. L., Eds., American Society for Microbiology, Washington, D.C., 1975, 465.
13. **Dring, G. J. and Gould, G. W.,** Sequence of events during rapid germination of spores of *Bacillus cereus*, *J. Gen. Microbiol.*, 65, 101, 1971.
14. **Pulvertaft, R. J. V. and Haynes, J. A.,** Adenosine and spore germination; phase-contrast studies, *J. Gen. Microbiol.*, 5, 657, 1951.
15. **Vary, J. C. and Halvorson, H. O.,** Kinetics of germination of *Bacillus* spores, *J. Bacteriol.*, 89, 1340, 1965.
16. **Hashimoto, T., Frieben, W. R., and Conti, S. F.,** Germination of single bacterial spores, *J. Bacteriol.*, 98, 1011, 1969.

17. **Hashimoto, T., Frieben, W. R., and Conti, S. F.,** Microgermination of *Bacillus cereus* spores, *J. Bacteriol.*, 100, 1385, 1969.
18. **Hashimoto, T., Frieben, W. R., and Conti, S. F.,** Kinetics of germination of heat-injured *Bacillus cereus* spores, in *Spores V,* Halvorson, H. O., Hanson, R., and Campbell, L. L., Eds., American Society for Microbiology, Washington, D.C., 1972, 409.
19. **Sogin, M. L., McCall, W. A., and Ordal, Z. J.,** Effect of heat activation conditions on the germinal response of *Bacillus cereus* T spores, in *Spores V,* Halvorson, H.O., Hanson, R., and Campbell, L. L., Eds., American Society for Microbiology, Washington, D.C., 1972, 363.
20. **Shacter, S. M. and Hashimoto, T.,** Bimodel kinetics of germination of *Bacillus cereus* T spores, in *Spores VI,* Gerhardt, P., Costilow, R. N., and Sadoff, H. L., Eds., American Society for Microbiology, Washington, D.C., 1975, 531.
21. **Hachisuka, Y., Asano, N., Kato, N., Okajima, M., Kitaori, M., and Kuno, T.,** Studies on spore germination I. Effect of nitrogen sources on spore germination, *J. Bacteriol.*, 69, 399, 1955.
22. **Woese, C. R., Morowitz, H. J., and Hutchinson, C. A., III,** Analysis of action of L-alanine analogues in spore germination, *J. Bacteriol.*, 76, 578, 1958.
23. **McCormick, N. G.,** Kinetics of spore germination, *J. Bacteriol.*, 89, 1180, 1965.
24. **Woese, C. R., Vary, J. C., and Halvorson, H. O.,** A kinetic model for bacterial spore germination, *Proc. Natl. Acad. Sci. U.S.A.*, 59, 869, 1969.
25. **Thibeault, D. and Lefebvre, G. M.,** Triggering in unactivated *Bacillus megaterium* spores, *Can. J. Microbiol.*, 30, 997, 1984.
26. **Lefebvre, G. M. and Antippa, A. F.,** The kinetics of germination in bacterial spores, *J. Theor. Biol.*, 95, 489, 1982.
27. **Vary, J. C.,** Properties of *Bacillus megaterium* temperature-sensitive germination mutants, *J. Bacteriol.*, 121, 197, 1975.
28. **Stewart, G. S. A. B., Johnstone, K., Hagleberg, E., and Ellar, D. J.,** Commitment of bacterial spores to germinate. A measure of the trigger reaction, *Biochem. J.*, 198, 101, 1981.
29. **Shibata, H., Takamatsu, H., and Tani, I.,** Germination of unactivated spores of *Bacillus cereus* T. Effect of preincubation with L-alanine or inosine on subsequent germination, *Jpn. J. Microbiol.*, 20, 529, 1976.
20. **Shibata, H., Takamatsu, H., Minami, M., and Tani, I.,** Inhibition by potassium ion of the pregerminative response to L-alanine of unactivated spores of *Bacillus cereus* T, *Microbiol. Immunol.*, 22, 443, 1978.
31. **Yousten, A. A.,** Germination of *Bacillus cereus* endospores: a proposed role for heat shock and nucleosides, *Can. J. Microbiol.*, 21, 1192, 1975.
32. **Leblanc, R. and Lefebvre, G. M.,** A stochastic model of bacterial spore germination, *Bull. Math. Biol.*, 46, 447, 1984.

Chapter 9

PHENOMENOLOGICAL THEORY FOR BACTERIAL CHEMOTAXIS

Gerald Rosen

TABLE OF CONTENTS

I. INTRODUCTION

As a widespread phenomenon in nature, many species of motile bacteria exhibit a chemotactic response toward oxygen, sugars, amino acids, and various other inorganic and organic substances. In particular, aerobic peritrichous bacteria such as *Escherichia coli* swim with a stochastic motion biased in the direction of increasing oxygen concentration in an aqueous medium where oxygen is at critically limiting concentrations relative to other vital substances. Since the swimming speeds of motile bacterial cells are less than or of the order of 10^{-2} cm/sec, the Reynolds number of their swimming motion is less than or of the order of 10^{-4}. Thus, micrometer-size motile bacterial are extremely small in a hydrodynamic sense, with their tactic swimming motion regulated by the viscosity of the surrounding aqueous medium. The relative flow of this medium is governed by the Navier-Stokes equations for viscous incompressible flow subject to zero relative-flow boundary conditions on the cell surfaces and uniform relative-flow at spatial infinity.

In cases for which the bacterial population is very large and the number of cells per unit volume varies in a virtually continuous fashion through the medium, one can formulate a phenomenological theory for bacterial chemotaxis. The distribution of chemotactic cells and the varying concentration of the limiting attractant are governed by coupled partial differential equations. Subject to initial and boundary conditions relevant to laboratory experiments, the coupled partial differential equations are generally amenable to exact or approximate analytical solution. This chapter reviews the mathematical details of the phenomenological theory for bacterial chemotaxis.

II. GOVERNING EQUATIONS

Quantitative laboratory experiments[1-13] have elucidated the swimming response of motile bacteria toward or away from various chemical attractants and repellants. Assuming that the bacterial cells act independently in their swimming motion and in their consumption of limiting substrate (which generally requires fewer than 10^9 cells/cm^3), the distribution of chemotactic cells and the concentration of the substrate are governed by the coupled partial differential equations:[14-39]

$$\partial n/\partial t = -\nabla \cdot \vec{f} \quad (1)$$

$$\vec{f} \equiv -\mu(s)\,\nabla n + \chi(s)\,n\nabla s \quad (2)$$

$$\partial s/\partial t = D\nabla^2 s - \kappa(s)\,n \quad (3)$$

where $n = n(\vec{x},t)$ denotes the number per unit volume of motile chemotactic cells in the neighborhood of the point $\vec{x}$ at time t; $s = s(\vec{x},t)$ denotes the concentration of the chemotactic agent; $\mu(s)$, $\chi(s)$, $\kappa(s)$ are the random motility, chemotactic flux, and consumption rate functions, respectively; and D denotes the constant diffusivity of the chemotactic agent. Equations 1 and 2 express the local conservation of the total number of cells subject to random motility and chemotactic flow (for inclusion and treatment of a population growth term see References 33, 35, and 37) while Equation 3 describes diffusion and bacterial consumption of the limiting substrate in the medium. Observe that the right side of the flux-vector Equation 2 is linear in n, and the total (random plus chemotactic) flux of cells defined by Equation 2 is the most general form being bilinear in n, with the gradient operator ∇ and with the coefficient functions $\mu(s)$ and $\chi(s)$ to be determined theoretically and/or empirically. In a similar fashion, the consumption rate function $\kappa(s)$ in Equation 3 must also be specified theoretically and/or empirically. Then the governing Equations 1, 2, and 3 constitute a deterministic system for predicting the forms of $n = n(\vec{x},t)$ and $s = s(\vec{x},t)$ for prescribed initial and boundary conditions.

III. RANDOM MOTILITY, CHEMOTACTIC FLUX, AND CONSUMPTION RATE FUNCTIONS

Because the cells are hydrodynamically small, the Navier-Stokes equation invariance symmetries[36] may show up in the tactic swimming motion over distances of the order of millimeters or centimeters and associated times of the order of minutes or hours, for on such a macroscopic migrational scale the cellular length and flagellar oscillation times of the bacteria are negligibly small in relative magnitude. The classical Navier-Stokes equations for three-dimensional incompressible viscous flow take the form

$$\partial\vec{u}/\partial t = -\vec{u}\cdot\nabla\vec{u} + \nu\nabla^2\vec{u} - \rho^{-1}\nabla p, \quad \nabla\cdot\vec{u} = 0 \tag{4}$$

where ν and ρ are positive constants. Equation 4 is invariant under the space-time dilatation transformations

$$\vec{x} \to \lambda\vec{x}, \quad t \to \lambda^2 t, \quad \vec{u} \to \lambda^{-1}\vec{u}, \quad p \to \lambda^{-2}p \tag{5}$$

where λ is a positive constant parameter. Under the space-time dilatations (Equation 5), the spatial gradient operator transforms as $\nabla \to \lambda^{-1}\nabla$ while the density variables in Equations 1 through 3 transform as

$$n \to \lambda^{-3}n, \quad s \to \lambda^{-3}s \tag{6}$$

Thus, Equations 1 through 3 feature the Navier-Stokes space-time dilatation invariance if and only if $\vec{f} \to \lambda^{-4}\vec{f}$ and

$$\mu(s) = \mu \equiv \text{(constant)} \tag{7}$$

$$\chi(s) = \delta s^{-1} \qquad \delta \equiv \text{(constant)} \tag{8}$$

$$\kappa(s) = \alpha_{2/3}s^{2/3} \qquad \alpha_{2/3} \equiv \text{(constant)} \tag{9}$$

Quantitatively accurate rapid-scanning densitometer experiments[13] have shown that Equations 7 and 8 do indeed apply to motile *Escherichia coli* cells attracted by low concentrations of oxygen under standard experimental conditions; the constant parameters in Equations 7 and 8 have the experimentally observed values[13]

$$\mu = [5.5(\pm\ 4.5)] \times 10^{-7}\ \text{cm}^2/\text{sec} \tag{10}$$

$$\delta = [1.8(\pm\ 0.7)]\ \mu \tag{11}$$

It is noteworthy that Equations 7 and 8 were proposed and employed originally[14] as the simplest expressions for the random motility and chemotactic flux functions which yield steadily propagating (traveling) band distributions of chemotactic bacteria; as pointed out here, Equations 7 and 8 are uniquely distinguished as the forms that engender invariance for Equations 1 and 2 under the Navier-Stokes space-time dilatations. In the case of a one-dimensional band of chemotactic bacteria that propagates in the +x direction with constant velocity u (see Section V), the cell density and limiting substrate concentration distributions are functions of the quantity $u(x - ut)$, an invariant under the transformations. Thus, the steadily propagating band solutions are space-time dilatation invariant.

A consumption rate function $\kappa(s)$ proportional to s^m, where the constant exponent m is positive and less than or equal to 1, has been suggested[18-21,25,27,31] on the basis of alternative theoretical considerations; more detailed transient measurements on chemotactic *E. coli* and/or other bacterial species are required to confirm the symmetry-associated $s^{2/3}$ dependence for the consumption rate function shown in Equation 9. It is well known that the rate functions for metabolic processes in biological species can exhibit similarity under dynamic as well as classical allometric scaling, and in particular, the principal metabolic processes and associated uptake of a critical substrate by a motile bacterium can manifest a dynamic scaling symmetry which relates directly to the viscosity-dominated swimming motion. In the case of motile *E. coli* cells in which the energy-yielding reactions are modulated by aerobic respiration, the limiting substrate will be oxygen if the local concentration is less than about 0.26 mg/ℓ at 37.8°C (decreasing to about 0.10 mg/ℓ at 15°C) and a metabolizable carbon-bearing nutrient is present in nonlimiting concentrations.[36] The latter critical oxygen concentration values are less than 4% of the oxygen-saturation concentrations for fresh water in contact with air at 1 atm, O_2 at 6.55 mg/ℓ at 37.8°C (increasing to 10.3 mg/ℓ at 15°C), and the oxygen consumption rate expression (Equation 9 may indeed supersede the Monod (i.e., Michaelis-Menten[40]) form

$$\kappa(s) = \kappa s/(s + \kappa_2)$$

at such small values for s. It is clear that experiments with very low oxygen concentrations are required to determine whether the rate of oxygen consumption has the functional dependence on s shown in Equation 9 in the case of motile *E. coli*. As shown in Table 1, many other motile bacterial species also exhibit chemotaxis toward or away from various substrates, with new chemotactic bacterial species being observed and added to the list each year.[41] An important question to be answered by future studies is whether Navier-Stokes dilatation symmetry in the phenomenological theory, as manifest in Equations 7 through 9, applies to diverse forms of bacterial chemotaxis at sufficiently low concentrations of limiting substrates. At moderate concentrations of such a substrate, one expects an s-range of validity for the linear consumption rate function,[17,18,29,34,35,38]

$$\kappa(s) = \alpha_1 s \qquad \alpha_1 = (\text{constant}) \tag{12}$$

while the Monod form for $\kappa(s)$ is expected to have a (higher) s-range of validity for substrates that continue to be limiting (i.e., metabolic process-controlling for the bacteria) relative to all other chemotactic nutrient substances in the medium at high values of s.

IV. ASSOCIATED FOKKER-PLANCK EQUATION AND THE THEORETICAL SIGNIFICANCE OF $\delta = 2\mu$

The substitution of Equations 2, 7, and 8 into Equation 1 yields

$$\frac{\partial n}{\partial t} = \mu \nabla^2 n - \delta \nabla \cdot (n s^{-1} \nabla s) \tag{13}$$

Table 1
CHEMOTACTIC RESPONSE OF MOTILE SPECIES OF BACTERIA[33]

Genus	Attractants	Repellents
Escherichia	Oxygen	pH extremes
	Sugars	Aliphatic alcohols
	Amino acids	
Pseudomonas	Oxygen	Inorganic ions
	Sugars	pH extremes
	Amino acids	Amino acids
	Nucleotides	
	Vitamins	
	Ammonium ions	
Bacillus	Oxygen	Inorganic ions
	Sugars	pH extremes
	Amino acids	Metabolic poisons
Salmonella	Oxygen	Aliphatic alcohols
	Sugars	
	Amino acids	
Vibrio	Oxygen	
	Amino acids	
Spirillum	Oxygen	Inorganic ions
	Sugars	pH extremes
	Amino acids	
Rhodospirillum	Nucleotides	pH extremes
	Sulfhydryl compounds	Poisons
Clostridium		Oxygen
Bdellovibrio	Amino acids	
Proteus	Oxygen	Inorganic acids
	Sugars	pH extremes
	Amino acids	
Erwinia	Sugars	Inorganic ions
		pH extremes
Sarcina		Inorganic ions
		pH extremes
Serratia	Oxygen	Inorganic ions
	Sugars	pH extremes
	Amino acids	
Bordetella	Oxygen	
Pasteurella	Oxygen	
Marine bacteria	Algal culture filtrates	Heavy metals
		Toxic hydrocarbons

If $N \equiv \int_R n\, d^3x$ denotes the constant total number of bacteria in a closed spatial region R with the normal components of ∇n and ∇s equal to zero over the boundary of R, then N is constant with time, and the normalized variable n/N is the probability density for finding a bacterium in the neighborhood of point $\vec{x}$ at time t. In terms of n/N, Equation 13 is the Fokker-Planck equation for stochastic motion of a single bacterium.[16] This Fokker-Planck equation follows directly and without approximation from the Langevin equation for bacterial swimming motion, the stochastic propulsive thrust of a bacterium being in dynamic equilibrium with its Stokes drag force during "run" intervals with the $d\vec{x}/dt$ constant.[16]

It has been pointed out[42,43] that certain Fokker-Planck equations can be transformed to associated canonical Schrodinger-Bloch equations for which there exists an extensive quantum statistical literature. In addition to being linear and parabolic, the distinguishing structural feature of a Schrodinger-Bloch equation is the absence of any first-order gradient term involving $\nabla\Psi$, where Ψ is the Schrodinger-Bloch dependent variable; spatial derivatives of Ψ come in exclusively through the second-order random diffusion term proportional to $\nabla^2\Psi$. Thus, for example, Equation 3 becomes the Schrodinger-Bloch equation

$$\frac{\partial s}{\partial t} = D\nabla^2 s - \alpha_1 ns \tag{14}$$

if $n = n(\vec{x},t)$ is as indicated above and the consumption rate function in Equation 3 is given by the linear expression Equation 12. In order to transform Equation 13 into its canonical Schrodinger-Bloch form, one must introduce the term,

$$\Psi \equiv n/s^{\delta/2\mu} \tag{15}$$

By computing the time derivative and Laplacian of Equation 15, it follows from Equation 13 that Equation 15 satisfies the Schrodinger-Bloch equation

$$\frac{\partial\Psi}{\partial t} = \mu\nabla^2\Psi - \phi\Psi \tag{16}$$

where the potential function, ϕ, is given by

$$\phi = \frac{1}{2}\mu^{-1}\delta\left[s^{-1}\left(\frac{\partial s}{\partial t} + \mu\nabla^2 s\right) + \left(\frac{1}{2}\delta - \mu\right)s^{-2}|\nabla s|^2\right] \tag{17}$$

It is interesting to note that the Schrodinger-Bloch function is invariant under the Navier-Stokes transformation if and only if the condition $\delta = 2\mu$ holds,

$$\Psi \rightarrow \Psi \quad \leftrightarrow \quad \delta = 2\mu \tag{18}$$

Therefore, the condition $\delta = 2\mu$, which is admitted by the experimental range indicated by Equation 11 for the low-concentration oxygen chemotaxis of motile *E. coli*, implies that the Schrodinger-Bloch function is dilatation invariant, according to Equation 18. Notice[34,36] also that the potential function is independent of first-order spatial derivatives in s if and only if $\delta = 2\mu$ and for this condition Equation 17 reduces to

$$\phi = s^{-1}\left(\frac{\partial s}{\partial t} + \mu\nabla^2 s\right) \tag{19}$$

Hence, the term involving the first-order spatial gradient ∇s is relinquished in the Schrodinger-Bloch equation if and only if $\delta = 2\mu$. With the conditon $\delta = 2\mu$, the Schrodinger-Bloch function becomes $\Psi = n/s$, the number of cells per unit mole of chemotactic agent in the neighborhood of $\vec{x}$ at time t. The resulting system of coupled Equations 3, 16, and 19 with $n = s\Psi$ has all the spatial symmetry admitted by the invariance group of the Laplacian operator ∇^2, unconstrained by the presence of first-order spatial derivative terms. In this sense, the condition $\delta = 2\mu$ reflects higher intrinsic symmetry in the governing equations, as well as dilatation invariance of the Schrodinger-Bloch function.

V. ANALYTICAL SOLUTIONS TO THE GOVERNING EQUATIONS

The following are representative examples of analytical solutions to the governing equations in the phenomenological theory of bacterial chemotaxis. These solutions describe basic, experimentally accessible, chemotactic distributions of bacteria and a consumed limiting substrate. For further details and experimental comparisons, the reader should consult the references that are cited below. Additional analytical solutions have also been reported in the literature.[7,18,21,27,28,30-32,35,38]

A. Steady-State Distributions of Bacteria Chemotactic Toward Oxygen Close to the Surface of an Oxygen-Depleted Medium[29]

Consider the one-dimensional steady-state solution to Equations 13 and 14 with z denoting distance into an oxygen-depleted aqueous medium from the z = 0 surface, where the number of motile chemotactic bacteria per unit volume and the oxygen concentration have the prescribed values

$$\left.\begin{array}{l} n = n_o \\ s = s_o \end{array}\right\} \quad \text{at } z = 0 \tag{20}$$

Suppose the bacterial cells and oxygen are absent far below the surface, so that

$$\left.\begin{array}{l} n = 0 \\ s = 0 \end{array}\right\} \quad \text{at } z = \infty \tag{21}$$

Specializing Equations 13 and 14 for n = n(z) and s = s(z), we have

$$\mu \frac{d^2n}{dz^2} - \delta \frac{d}{dz}\left(ns^{-1}\frac{ds}{dz}\right) = 0 \tag{22}$$

$$D\frac{d^2s}{dz^2} - \alpha_1 ns = 0 \tag{23}$$

The exact solution to Equations 22 and 23 subject to the boundary conditions shown in Equations 20 and 21 is given by the remarkable simple closed-form expressions

$$n = n_o[1 + (z/z_*)]^{-2} \tag{24}$$

$$s = s_o[1 + (z/z_*)]^{-2\mu\delta^{-1}} \tag{25}$$

in which the characteristic thickness of the bacterial congregation and oxygen-containing layer appears as

$$z_* = [(D\alpha_1^{-1}n_o^{-1})(2\mu\delta^{-1})(1 + 2\mu\delta^{-1})]^{1/2} \tag{26}$$

Notice that the thickness formula (Equation 26) is wholly independent of s_0 and depends on the motility and chemotactic flux coefficient exclusively through the dimensionless ratio $2\mu\delta^{-1}$; thus, for $\delta = 2\mu$, Equation 26 reduces to $z_* = (2D/\alpha_1 n_0)^{1/2}$. The total number of cells under a unit surface area follows from the solution function (Equation 24) as

$$\mathcal{N} = \int_0^\infty n(z)\,dz = n_o \int_0^\infty [1 + (z/z_*)]^{-2}\,dz = n_o z_* \tag{27}$$

and hence, by using Equation 26 one obtains

$$\mathcal{N} z_* = D\alpha_l^{-1}(2\mu\delta^{-1})\,(1 + 2\mu\delta^{-1}) \tag{28}$$

The result expressed by Equation 28 shows that the thickness z_* varies inversely with the total number of cells under a unit surface area, if conditions influencing the consumption constant α_l and the dimensionless ratio $2\mu\delta^{-1}$ are held fixed. It has been shown[38] that the solution (Equations 24 and 25) is stable with respect to arbitrary perturbations in n and s; i.e., the dynamical Equations 13 and 14 require perturbed forms for n and s to return to those given by Equations 24 and 25 as t increases.

B. Steadily Propagating One-Dimensional Bands[14,18]

The consumption rate functions (Equations 9 and 12) are subsumed by the more general form

$$\kappa(s) = \alpha_m s^m \qquad m = (\text{constant}) \tag{29}$$

$$\alpha_m = (\text{constant})$$

which reduces to the Navier-Stokes symmetric expression (Equation 9) for m = 2/3 and the linear-theoretic expression (Equation 12) for m = 1. Let us consider the specialization of Equation 3 obtained by setting D = 0 and representing the consumption rate function by Equation 29:

$$\partial s/\partial t = -\alpha_m s^m n \tag{30}$$

In combination with Equation 13, Equation 30 admits exact analytical solutions depending on the dimensionless variable

$$\xi \equiv \mu^{-1}u(x - ut) \tag{31}$$

in which the propagation velocity, μ, is a positive constant parameter. These steadily propagating one-dimensional "traveling" band solutions take the form

$$n = n(\xi) = (u^2 s_\infty^{1-m}/\alpha_m\gamma)\,e^{-\xi}(1 + e^{-\xi})^{-\delta/\gamma} \tag{32}$$

$$s = s(\xi) = s_\infty(1 + e^{-\xi})^{-\mu/\gamma} \tag{33}$$

with

$$\left.\begin{aligned} n &= 0 \\ s &= 0 \end{aligned}\right\} \text{ at } \xi = -\infty \tag{34}$$

and

$$\left.\begin{aligned} n &= 0 \\ s &= s_\infty \end{aligned}\right\} \quad \text{at } \xi = +\infty \tag{35}$$

provided that the quantity $\gamma = \delta - (1 - m)\,\mu$ is in the range $0 < \gamma < \delta$, or equivalently if

$$[1 - (\delta/\mu)] < m < 1 \tag{36}$$

The latter necessary and sufficient condition for steadily propagating one-dimensional bands is generally satisfied by the Navier-Stokes symmetric case $m = 2/3$, but such "traveling" band solutions are precluded by Equation 36 for the linear-theoretic $m = 1$ case. The total number of cells per unit cross-sectional area of band is

$$\mathcal{N} \equiv \int_{-\infty}^{\infty} n\,dx = \mu u^{-1} \int_{-\infty}^{\infty} n(\xi)\,d\xi = [u s_\infty^{1-m}/(1 - m)\,\alpha_m] \tag{37}$$

where Equation 32 had been integrated explicitly. According to Equation 37, the velocity of propagation u and the total number of cells per unit cross-sectional area, $\mathcal{N}$, are directly proportional in this one-parameter family of steadily propagating bands; the solutions can be viewed as labeled unambiguously by either u or $\mathcal{N}$ for initially given values of m, α_m, δ, μ, and s_∞.

It has been demonstrated by elaborate mathematical analysis[20,22] that the band solutions (Equations 32 and 33) are dynamically stable with respect to arbitrary perturbations governed by Equations 13 and 30 for all u (>0). Also, for Equations 1 through 3 with $D = 0$ and $\mu(s)$, $\chi(s)$, $\kappa(s)$ unspecified at the outset, it has been shown[24,26] that the existence of a steadily propagating band solution puts certain necessary and sufficient conditions on the functional form of $\mu(s)$, $\chi(s)$, $\kappa(s)$; the latter conditions reduce to Equation 36 if one invokes Equations 7, 8, and 29.

C. Time-Dependent Formation of One-Dimensional Bands[35]

Let us now consider the one-dimensional forms of Equations 3 and 13 using Equation 29,

$$\frac{\partial n}{\partial t} = \mu \frac{\partial^2 n}{\partial x^2} - \delta \frac{\partial}{\partial x}\left(n s^{-1} \frac{\partial s}{\partial x}\right) \tag{38}$$

$$\frac{\partial s}{\partial t} = D \frac{\partial^2 s}{\partial x^2} - \alpha_m s^m n \tag{39}$$

subject to the boundary conditions for all $t \geqq 0$

$$\left.\begin{aligned} \partial n/\partial x &= 0 \\ \partial s/\partial x &= 0 \end{aligned}\right\} \quad \text{at } x = 0 \tag{40}$$

$$\left.\begin{aligned} n &= 0 \\ s &= s_\infty \end{aligned}\right\} \quad \text{at } x = \infty \tag{41}$$

and the initial conditions at t = 0,

$$n(x,0) = \begin{cases} n_o(\equiv \text{const}) & \text{for } 0 \leqq x \leqq x_o \\ 0 & \text{for } x > x_o \end{cases} \tag{42}$$

$$s(x,0) = s_\infty \quad \text{for } 0 \leqq x < \infty \tag{43}$$

Observe that Equation 42 describes a uniform distribution of cells in the interval $0 \leqslant x \leqslant x_0$ at $t = 0$. Capillary tube experiments[1-3] which feature the formation and eventual steady propagation of "traveling" bands of bacteria are modeled by Equations 38 through 43. In the analytical iteration solution[35] to this initial-value/boundary-value problem, a band evolves dynamically and breaks free from a residual concentration of cells in the interval $0 \leqslant x \gtrsim x_0$. The asymptotically constant propagation velocity of the band is given by the formula

$$u = (\delta\alpha_m s_\infty^{m-1} n_o)^{1/2} \tag{44}$$

in the "diffusionless" first approximation, valid for $\delta >> \max\{\mu, D\}$.

Thus, the unique propagation velocity given by Equation 44 emerges for the steadily propagating band, with the total number of cells per unit cross-sectional area given in turn by Equations 37 and 44. For refined formulas produced by the second iterative approximation, which brings in μ and D, the reader should consult the original paper.[35]

VI. APPLICATION OF THE SOLUTIONS TO SYMBIOTIC BACTERIAL ASSOCIATIONS IN NATURE

The chemotaxis of motile bacteria plays a key role in symbiotic associations of microorganisms in natural habitats,[41,44] and the phenomenological theory for bacterial chemotaxis can be applied to illuminate certain quantitative aspects of these symbiotic associations. An interesting practical example is provided by the symbiotic association of methane-producing and motile aerobic bacteria in certain fresh water habitats, for which the phenomenological theory can be applied as follows.[39]

Oxygen-depleted, organically rich ponds, swamps, and other fresh water habitats provide natural ecosystems for large steady-state populations of methanobacteria.[45] Experimental measurements of the methane flux from various habitats[46-49] show that the methanobacteria can prosper in several qualitatively different types of microbial ecosystems. In particular, for certain ponds and swamps the methane flux, f, is given to an accuracy of about 10% by the simple empirical fomula[47]

$$f = (4.1\,T - 42)\ \mu g/cm^2/day \tag{45}$$

where T is the temperature in degrees Celsius through the experimental range 12°C < T < 26°C. It is noteworthy that Equation 45 applies to aquatic habitats of various depths and sediment compositions, even though no parametric dependence on the latter variables is evident. For other habitats the methane flux has been observed experimentally to vary with depth and sediment composition,[48,49] and Equation 45 is not applicable in such cases.

Since Equation 45 applies to fresh water habitats of various depths and sediment compositions, methane production must be rate controlled by biochemical and biophysical processes which take place near the water-atmosphere interface. Equation 45 is quantitatively consistent with a microbial ecosystem in which O_2-chemotactic, motile, aerobic bacteria supply the CO_2 for anaerobic respiration of methanobacteria. Commensal with the O_2-shielding and CO_2-providing motile aerobic bacteria, the methanobacteria liberate CH_4 at precisely the molal rate that oxygen is consumed and CO_2 is generated by aerobic respiration of the former:

Aerobic bacteria:

$$O_2 + (\text{carbon-bearing nutrient}) \rightarrow CO_2 + (\text{other metabolites}) \tag{46}$$

Methanobacteria:

$$CO_2 + 2C_2H_5OH \rightarrow CH_4 + 2CH_3COOH \tag{47}$$

Close to the surface of the oxygen-depleted body of water, the steady-state distributions of O_2-chemotactic motile aerobic bacteria and of dissolved oxygen are given by Equations 24 and 25 with Equation 26. To evaluate the thickness parameter, first observe that

$$D = 7.6 \times 10^{-2}\ \text{cm}^2/\text{hr} \tag{48}$$

is the oxygen diffusivity constant at temperatures of biological interest and

$$\alpha_1 = (2.5 \times 10^{-8}/\text{cm}^3\ \text{cell/hr})\ \phi(T) \tag{49}$$

is a reasonable estimate for the fractional rate of O_2 consumption per unit concentration of bacterial cells. The dimensionless temperature-dependent factor $\phi(T)$ in Equation 49 relates the fact that O_2 consumption increases with increasing T, while the constant factor in Equation 49 derives from experimental data[35] on motile, aerobic, O_2-chemotactic *Escherichia coli,* for which

$$\phi(20) \cong 1 \qquad d\phi/dT|_{T=20} > 0 \tag{50}$$

If one assumes that the bacterial cells and oxygen concentration have the typical surface values

$$n_o = 2.0 \times 10^8\ \text{cells/cm}^3 \tag{51}$$

$$s_o = 8.0\ \mu\text{g/cm}^3 \tag{52}$$

and supposes that $\delta = 2\mu$, then Equations 24 and 25 become

$$n = n_o[1 + (z/z_*)]^{-2} \tag{53}$$

$$s = s_o[1 + (z/z_*)]^{-1} \tag{54}$$

while Equation 26 reduces to

$$z_* = (0.174\ \text{cm})[\phi(T)]^{-1/2} \tag{55}$$

in view of Equations 48, 49, and 51. The total rate of O_2 consumption by the motile aerobic population is

$$\begin{pmatrix} O_2 \text{ consumption} \\ \text{rate per unit} \\ \text{surface area} \end{pmatrix} = \int_0^{\infty} \alpha_1 ns\, dz = -D(ds/dz)_{z=0}$$

$$= D\, z_*^{-1}\, s_o = (3.5\ \mu g/cm^2/hr)[\phi(T)]^{1/2} \qquad (56)$$

by virtue of Equations 23, 48, 52, 54, and 55. Since each mole of O_2 produces 1 mol of CO_2 through aerobic bacterial respiration according to Equation 46, and each mole of CO_2 is reduced to 1 mol of CH_4 by commensal methanogenic anaerobes according to Equation 47, the associated methane flux is given by 1/2 (= mol wt of CH_4/ mol wt of O_2) of the O_2-consumption rate per unit surface area shown in Equation 56, i.e.,

$$f = \frac{1}{2} D z_*^{-1}\, s_o = (42\ \mu g/cm^2/day)[\phi(T)]^{1/2} \qquad (57)$$

where the unit of time is taken as 1 day in order to facilitate comparison with Equation 45. The theoretical methane-flux expression (Equation 57) agrees with the experimental formula (Equation 45) if

$$\phi(T) \cong [(T/10) - 1]^2 \qquad (58)$$

and indeed the latter form (Equation 58) is in accord with the normalization and monotonicity conditions in Equation 50. Thus, for motile-aerobic bacteria with an O_2-consumption rate given by Equations 50 and 56, the typical parameter values shown in Equations 49, 51, and 52 yield the methane flux expression of Equation 45 as a consequence of commensalism between the aerobic and methanobacteria.

REFERENCES

1. **Adler, J.**, Chemotaxis in bacteria, *Science*, 153, 708, 1966.
2. **Adler, J.**, Effect of amino acids and oxygen on chemotaxis in *Escherichia coli*, *J. Bacteriol.*, 92, 121, 1966.
3. **Hazelbauer, G. L., Meshbov, R. E., and Adler, J.**, *Escherichia coli* mutants defective in chemotaxis toward specific chemicals, *Proc. Natl. Acad. Sci. U.S.A.*, 64, 1300, 1969.
4. **Berg, H. C. and Brown, D. A.**, Chemotaxis in *Escherichia coli* analysed in three-dimensional tracking, *Nature (London)*, 239, 500, 1972.
5. **Dahlquist, F. W., Lovely, P., and Koshland, D. E.**, Quantitative analysis of bacterial migration in chemotaxis, *Nature (London)*, 236, 120, 1972.
6. **Macnab, R. M. and Koshland, D. E.**, The gradient sensing mechanism in bacterial chemotaxis, *Proc. Natl. Acad. Sci. U.S.A.*, 69, 2509, 1972.
7. **Nossal, R.**, Growth and movement of rings of chemotactic bacteria, *Exp. Cell Res.*, 75, 138, 1972.
8. **Nossal, R. and Chen, S. H.**, Light scattering from motile bacteria, *J. Phys.*, 33, 12, 1972.
9. **Nossal, R. and Chen, S. H.**, Effects of chemoattractants on the motility of *Escherichia coli*, *Nature (London)*, 244, 253, 1973.
10. **Chapman, P.**, Chemotaxis in Bacteria, Ph.D. thesis, University of Minnesota, Minneapolis, 1973.
11. **Lovely, P. S. and Dahlquist, F. W.**, Statistical measures of bacterial motility and chemotaxis, *J. Theor. Biol.*, 50, 477, 1975.
12. **Adler, J.**, Chemotaxis in bacteria, *Ann. Rev. Biochem.*, 44, 344, 1975.
13. **Holz, M. and Chen, S. H.**, Spatio-temporal structure of migrating chemotactic band of *Escherichia coli*, *Biophys. J.*, 26, 243, 1979.
14. **Keller, E. F. and Segel, L. A.**, Traveling bands of chemotactic bacteria: a theoretical analysis, *J. Theor. Biol.*, 30, 235, 1971.

15. **Segel, L. A. and Stoeckly, B.,** Instability of a layer of chemotactic cells, attractant and degrading enzyme, *J. Theor. Biol.*, 36, 561, 1972.
16. **Rosen, G.,** Fundamental theoretical aspects of bacterial chemotaxis, *J. Theor. Biol.*, 41, 201, 1973.
17. **Segel, L. A. and Jackson, J. L.,** Theoretical analysis of chemotactic movement in bacteria, *J. Mechanochem. Cell Motil.*, 2, 25, 1973.
18. **Rosen, G.,** On the propagation theory for bands of chemotactic bacteria, *Math. Biosci.*, 20, 185, 1974.
19. **Rosen, G.,** Propagation theory for tactic bacteria — addendum, *Math. Biosci.*, 52, 303, 1980.
20. **Baloga, S. M.,** On the Stability and Structure of Steadily Propagating Bands and Rings of Chemotactic Bacteria, Ph.D. thesis, Drexel University, Philadelphia, 1974.
21. **Rosen, G.,** Bacterial chemotaxis in the temporal gradient apparatus, *Math. Biosci.*, 24, 17, 1975.
22. **Rosen, G.,** On the stability of steadily propagating bands of chemotactic bacteria, *Math. Biosci.*, 24, 273, 1975.
23. **Rubin, P. E.,** Form of the Flux Coefficient in the Theory for Bacterial Chemotaxis, Ph.D. thesis, Drexel University, Philadelphia, 1975.
24. **Keller, E. F. and Odell, G.,** Necessary and sufficient conditions for chemotactic bands, *Math. Biosci.*, 27, 309, 1975.
25. **Rosen, G.,** Analytical solution to the initial-value problem for traveling bands of chemotactic bacteria, *J. Theor. Biol.*, 49, 311, 1975.
26. **Rosen, G.,** Existence and nature of band solutions to generic chemotactic transport equations, *J. Theor. Biol.*, 59, 243, 1976.
27. **Rosen, G. and Baloga, S.,** On the structure of steadily propagating rings of chemotactic bacteria, *J. Mechanochem. Cell Motil.*, 3, 225, 1976.
28. **Lapidus, I. R. and Schiller, R.,** Model for the chemotactic response of a bacterial population, *Biophys. J.*, 16, 779, 1976.
29. **Rosen, G.,** Steady-state distribution of bacteria chemotactic toward oxygen, *Bull. Math. Biol.*, 40, 671, 1978.
30. **Lapidus, I. R. and Schiller, R.,** A model for traveling bands of chemotactic bacteria, *Biophys. J.*, 22, 1, 1978.
31. **Lonngren, K. E. and Johnson, S. F.,** On a coupled set of nonlinear diffusion equations, *J. Appl. Phys.*, 51, 5201, 1980.
32. **Lapidus, I. R.,** Analysis of bacterial chemotaxis in flowing water, *Math. Biosci.*, 54, 79, 1981.
33. **Lauffenburger, D., Aris, R., and Keller, K.,** Effects of cell motility and chemotaxis on microbial population growth, *Biophys. J.*, 40, 209, 1982.
34. **Rosen, G.,** Theoretical significance of the condition $\delta = 2\mu$ in bacterial chemotaxis, *Bull. Math. Biol.*, 45, 151, 1983.
35. **Rosen, G.,** Analytical solutions for distributions of chemotactic bacteria, *Bull. Math. Biol.*, 45, 837, 1983.
36. **Rosen, G.,** Navier-Stokes symmetry in the phenomenological transport theory for bacterial chemotaxis, *Phys. Rev. A*, 29, 2774, 1984.
37. **Lauffenburger, D., Kennedy, C. R., and Aris, R.,** Traveling bands of motile bacteria in the context of population survival, *Bull. Math. Biol.*, 46, 19, 1984.
38. **Rosen, G.,** Stability of spatially heterogeneous steady-state distributions of oxygen-chemotactic aerobic/anaerobic bacteria, *Bull. Math. Biol.*, 46, 235, 1984.
39. **Rosen, G.,** Commensalism of methane-producing and motile aerobic bacteria in certain freshwater wetlands, *Bull. Math. Biol.*, 46, 333, 1984.
40. **Murray, J. D.,** *Lectures on Nonlinear Differential Equation Models in Biology*, Clarendon Press, Oxford, 1977, 2.
41. **Gallucci, K. K. and Paerl, H. W.,** *Pseudomonas aeruginosa* chemotaxis associated with blooms of N_2-fixing blue-green algae *(Cyanobacteria)*, *Appl. Environ. Microbiol.*, 45, 557, 1983.
42. **Goel, N. C., Maitra, S. C., and Montroll, E. W.,** On the Volterra and other nonlinear models of interacting populations, *Rev. Mod. Phys.*, 43, 231, 1971.
43. **Montroll, E. W.,** Some statistical aspects of the theory of interacting species, in *Some Mathematical Problems in Biology III*, American Mathematics Society, Providence, R. I., 1972, 101.
44. **Jones, J. G.,** Studies on freshwater bacteria factors which influence the population and its activity, *J. Ecol.*, 59, 593, 1971.
45. **Zeikus, J. G.,** Metabolic communication between biodegradative populations in nature, *Microbes in their Natural Environments*, Cambridge University Press, London, 1983, 423.
46. **Koyama, T.,** Gaseous metabolism in lake sediments and paddy soils and the production of atmospheric methane and hydrogen, *J. Geophys. Res.*, 68, 3971, 1963.
47. **Baker-Blocker, A., Donahue, T. M., and Mancy, K. H.,** Methane flux from wetland areas, *Tellus*, 29, 245, 1977.

48. **Harriss, R. C. and Sebacher, D. I.,** Methane flux in forested freshwater swamps of the southeastern United States, *Geophys. Res. Lett.*, 8, 1002, 1981.
49. **Sebacher, D. I. and Harriss, R. C.,** A system for measuring methane fluxes from wetland environments, *J. Environ. Qual.*, 11, 34, 1982.

Chapter 10

ATTACHMENT MECHANISMS IN THE SURFACE GROWTH OF MICROORGANISMS

Paul R. Rutter and Brian Vincent

TABLE OF CONTENTS

I. INTRODUCTION

The attachment of microorganisms to surfaces is one, albeit complex, example of the general problem of particle "adsorption" on substrates. The questions one seeks to answer in this type of problem are

1. What is the net rate of deposition of particles per unit area of the substratum?
2. What is the final (equilibrium or steady state) number of particles per unit area?

Clearly answers to these questions depend upon a number of conditions. These may be summarized as follows:

1. An open or closed system, i.e., is the total number of particles available in the system constant or variable?
2. The local hydrodynamic conditions in the vicinity of the substratum; e.g., is the fluid phase stationary or flowing? Is the flow laminar or turbulent?
3. Local interaction forces between the particles and substrate, and their possible time dependence.
4. Masking effects due to previously adsorbed particles.
5. Surface heterogeneity: physical (e.g., roughness) or chemical.
6. Preaggregation in the fluid phase.
7. In the case of microorganisms one has to include cell growth and division in the fluid phase and/or on the substratum surface.

In this review we shall restrict discussion to points (1) to (4): points (5) to (7) are difficult to incorporate, in a quantitative way, into models of microorganism adsorption, particularly if there are variations in the strength of attachment between primary adsorbed cells and their daughters.

We shall start Section II by considering the simplest case: i.e., the adsorption of microorganisms from, first, a closed system, and then open systems under stationary fluid conditions; we then introduce effects due to fluid flow (Section III). A fuller review of the subject has been given recently by Adamczyk et al.[1]

II. ADSORPTION UNDER STATIONARY FLUID CONDITIONS

A. General Analysis of Adsorption Rates in Closed Systems

In a closed system (total number of microorganisms fixed) one may monitor either the number of adsorbed microorganisms per unit area (Γ) with time, or the change in the number concentration of particles in the fluid phase, C. Γ and C are linked through the equation,

$$\Gamma = (C_o - C)\, V/S \tag{1}$$

Where C_o is the initial microorganism concentration, V is the total volume of suspension, and S the total surface area of the substratum. Alternatively,

$$\theta = \frac{\Gamma}{\Gamma_{max}} = \frac{(C_o - C)\, V}{\Gamma_{max} S} = (C_o - C)\, V/n \tag{2}$$

where θ = coverage, n = number of surface sites, and $\Gamma_{max} = n/S$

Thus, one may directly follow the net adsorption rate, i.e.,

$$d\Gamma/dt = J(1 - \theta) - \overleftarrow{k}\,\Gamma \tag{3}$$

where J is the flux toward the substratum surface, and $\overleftarrow{k}$ is the desorption rate constant, or

$$d\theta/dt = \frac{SJ(1 - \theta)}{n} - \overleftarrow{k}\,\theta \tag{4}$$

$$\equiv \overrightarrow{k}\,C(1 - \theta) - \overleftarrow{k}\,\theta \tag{5}$$

where $\overrightarrow{k}$ is the adsorption rate constant.

Alternatively, one may follow dC/dt. From Equations 1 to 5:

$$dC/dt = -\overrightarrow{k}\,nC(1 - \theta)/V + \overleftarrow{k}(C_o - C) \tag{6}$$

$$= -\overrightarrow{k}\,C\left[\frac{n}{V} - (C_o - C)\right] + \overleftarrow{k}(C_o - C) \tag{7}$$

Note that putting $d\theta/dt = 0$ in Equation 5, leads to the Langmuir adsorption isotherm,[2] i.e.,

$$\frac{\theta_{eq}}{1 - \theta_{eq}} = \frac{\overrightarrow{k}}{\overleftarrow{k}}\,C_{eq} = KC_{eq} \tag{8}$$

where θ_{eq} is the equilibrium coverage and C_{eq} the equilibrium concentration of microorganisms in the bulk phase.

Equation 7 may be integrated in various ways.[3,4] One integrated form (in which $\overleftarrow{k}$ is eliminated by introducing Equation 8) is,[4]

$$y = \frac{\overrightarrow{k}\,n}{V}\left[\frac{C_o}{C_o - C_{eq}} - \theta_{eq}\right]t$$

$$\text{where} \quad y = -\log_e\left\{\frac{C - C_{eq}}{C_o - C_{eq}} \Big/ [1 - (1 - C/C_o)\,\theta_{eq}]\right\} \tag{9}$$

Hence, $\overrightarrow{k}$ and $\overleftarrow{k}$ (and hence K) may be calculated from experimental C(t) data.

Vincent et al.[4] have explored the application of Equation 9 to the adsorption of small (0.2 μm diameter), positive polystyrene latex particles onto much larger (2 μm diameter), negative polystyrene latex beads and found that plots of y against t are indeed linear and pass through the origin; reasonable agreement was found between the value of $\overrightarrow{k}$ derived from the slopes of such plots and theoretical values of $\overrightarrow{k}$, based on the Smoluchowski theory of coagulation kinetics,[5] at least in the limit $t \rightarrow 0$ (i.e., low coverage).

B. Microorganism/Substratum Interactions

Theoretical evaluation of $\overrightarrow{k}$ and $\overleftarrow{k}$ requires knowledge of the microorganism/substratum interactions, and at higher coverages, the microorganism/microorganism interactions, since lateral interactions between adsorbed microorganisms will then also play a role. One may consider these forces to fall into three classes: dispersive, electrostatic, structural.

Several expressions for the ubiquitous *dispersion* interactions exist. For plane/sphere geometry the interaction energy, $V_d(h)$ is given by:

$$V_d(h) = -\frac{A}{12}\left[\frac{2x + 1}{x(x + 1)} + 2\log_e\left(\frac{x}{x + 1}\right)\right] \simeq -\frac{A}{12x}, \quad \text{when } x \ll 1 \tag{10}$$

for two spheres of equal size;

$$V_d(h) = -\frac{A}{12}\left[\frac{1}{x^2 + 2x} + \frac{1}{x^2 + 2x + 1} + 2\log_e\left(\frac{x^2 + 2x}{x^2 + 2x + 1}\right)\right]$$

$$\simeq -\frac{A}{24x}, \quad \text{when } x \ll 1 \tag{11}$$

where A is the composite Hamaker constant, given by

$$A = (A_1^{1/2} - A_2^{1/2})(A_3^{1/2} - A_2^{1/2}), \quad \text{and } x = h/2a \tag{12}$$

where a is the particle radius and h the distance between the particle and substratum surfaces; "1" refers to the substratum, "2" to the suspending medium (e.g., water), and "3" to the particle.

Hogg et al.[7] have derived a corresponding expression for the *electrostatic* interaction, V_e^Ψ and V_e^σ(at constant potential and constant charge, respectively) for plane/sphere geometry;

$$V_e^\Psi(h) = \pi\epsilon\epsilon_0 a(\Psi_s^2 + \Psi_p^2)\left[\frac{2\Psi_s\Psi_p}{\Psi_s^2 + \Psi_p^2}\log\left(\frac{1 + \exp\kappa h}{1 - \exp\kappa h}\right) + \log_e(1 - \exp(-2\kappa h))\right] \tag{13}$$

$$V_e^\sigma(h) = \pi\epsilon\epsilon_0 a(\Psi_s^2 + \Psi_p^2)\left[\frac{2\Psi_s\Psi_p}{\Psi_s^2 + \Psi_p^2}\log\left(\frac{1 + \exp\kappa h}{1 - \exp\kappa h}\right) - \log_e(1 - \exp(-2\kappa h))\right] \tag{14}$$

for two spheres of equal size;

$$V_e^\sigma(h) = -2\pi\epsilon\epsilon_0 a\, \Psi_s^2\log_e(1 - \exp(-\kappa h)) \tag{15}$$

$$V_e^\Psi(h) = -2\pi\epsilon\epsilon_0 a\, \Psi_s^2\log_e(1 + \exp(-\kappa h)) \tag{16}$$

where for small double layer overlap exp $(-\kappa h) \ll 1$ Equations 15 and 16 reduce to:

$$V_e(h) = 2\pi\epsilon\epsilon_0 a\, \Psi_s^2\exp(-\kappa h) \tag{17}$$

where ϵ is the dielectric constant of the medium (78.2 for water at 25°C), ϵ_0 is the permittivity of free space (8.85×10^{-12}J/C^2/m), and Ψ_p and Ψ_s are the electrical potentials at the particle and substratum surfaces. κ depends upon the total ion concentration in the bulk solution and is given by

$$\kappa^2 = \frac{N_A e^2}{\epsilon\epsilon_0 kT}\sum_i z_i C_i \tag{18}$$

where C_i is the concentration and z_i is the valency of the ions of type i, N_A is the Avogadro constant, k is the Boltzman constant, e is the electron charge, and T is the absolute temperature.

In the case of univalent electrolytes in aqueous solution at 25°C the double layer thickness (κ^{-1}) is related to the electrolyte concentration, c, as follows;

c/(mol/dm³)	**nm**
10^{-5}	100
10^{-3}	10
10^{-1}	1

Note 1: In Equations 13 and 14, $V_e \neq 0$ if either Ψ_s or $\Psi_p = 0$, (i.e., there may still be an electrostatic repulsion or attraction, for constant charge and constant potential, respectively) even if one of the surfaces is at zero potential.

Note 2: Since Ψ is not directly amenable to experimental determination, it is common practice to use the zeta potential as derived from electrokinetic experiments (e.g., electrophoresis).

Structural forces are associated with the change in the structure of the two interfacial regions surrounding the microorganism and the substratum, brought about by their overlap. These may be repulsive or attractive. For example, in aqueous media, the structure of the interfacial region near a surface will be different from bulk water. Close approach of two surfaces therefore results in displacement of the "interfacial water" into the bulk. For two hydrophilic surfaces this results in an extra repulsive force; for two hydrophobic surfaces an extra attractive force results; for a hydrophilic in conjunction with a hydrophobic surface the result is not obvious.

Similarly, structural forces occur when adsorbed (neutral) polymer is present on the microorganism and/or the substratum. When the two surfaces approach, perturbation of the polymer conformations occurs (in extreme cases polymer being displaced into the bulk solution). Again, the resultant forces induced may be repulsive (particularly at high coverage) or attractive (low coverage). It would be inappropriate to discuss this topic in detail here (extensive reviews have recently been given by Vincent and Whittington[8] and by Napper[9]); mention should also be made of the considerable advance in the interpretation of polymer-induced effects in colloid stability resulting from the advent of the Scheutjens-Fleer[10] theory of polymers at interfaces. We will just highlight three aspects which are relevant to the subject of particle adhesion:

1. If both the particle and the substrate carry adsorbed layers of neutral polymers (thickness δ_p and δ_s, respectively) and the polymers are in a "good" solvent environment, then a steeply rising ("steric") force operates at $h < (\delta_p + \delta_s)$, indeed for current purposes we may let $V_s = \infty$ for this condition (See Figure 1A).
2. If only one of the surfaces carries an adsorbed layer then an attractive force may be generated due to "bridging" of the polymer chain between the particle and substrate surfaces (See Figure 1B). This attractive interaction is difficult to quantify in a simplistic way, but clearly the net interaction (χ_s) of the polymer segments with the surface is important. The importance of polymer bridging between microorganisms and substratum surfaces has been suggested by a number of authors, both from theoretical

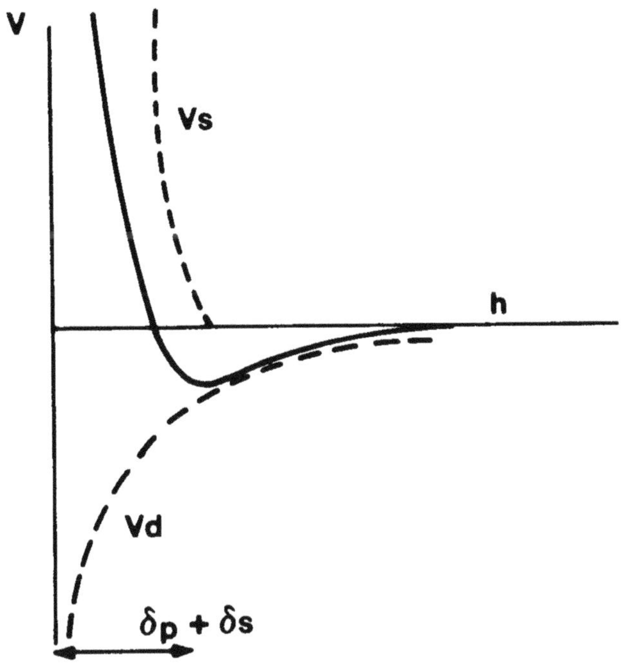

A

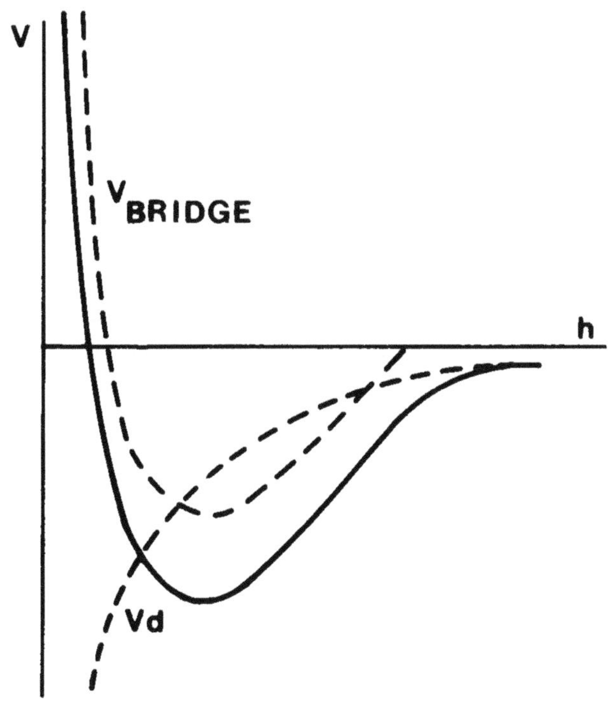

B

FIGURE 1. The effect of adsorbed polymer layers on the adsorption of microorganisms to surfaces.

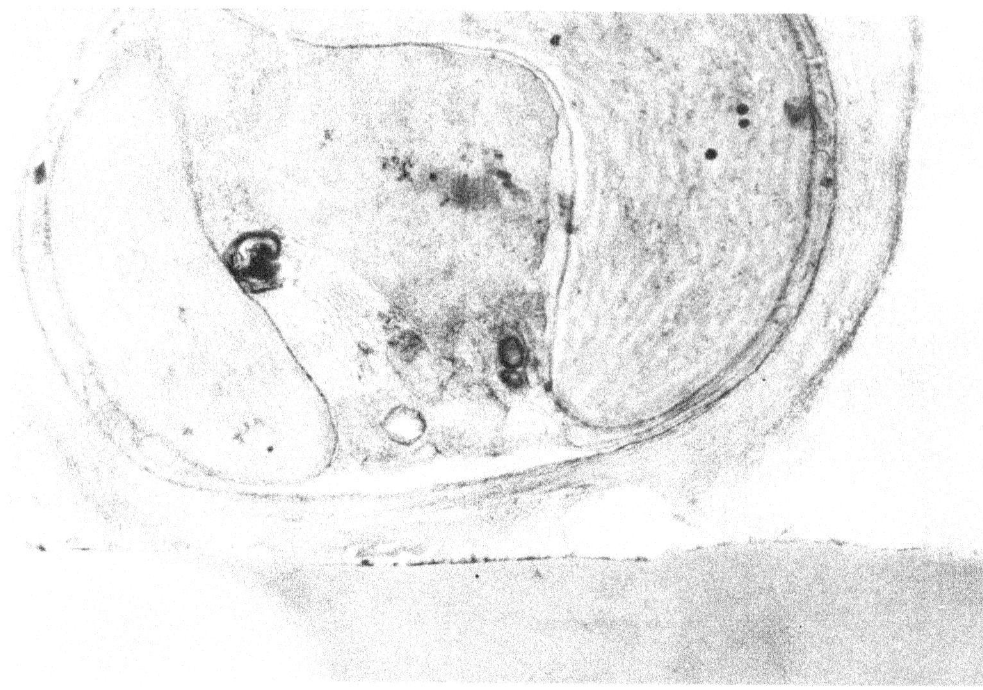

FIGURE 2. A transverse section of the prostrate vegetative system of an ectocarpalean alga attached to a plastic surface by means of extracellular polymer. (Photograph taken by Dr. A. H. L. Chamberlain, Surrey University.)

considerations[11-12] and because of the ubiquitous presence of polymer material interposed between the microorganism and substratum surface[13] observed in electron micrographs (see Figure 2).

3. If neither surface has any adsorbed polymer, but there is free polymer in solution (i.e., nonadsorbed), then an additional attractive force may result (V_{dep}) due to depletion effects.[14] The depletion zone near a nonadsorbing surface ($\chi_s < \chi_{s,crit}$) in a polymer solution has a lower segment concentration than the bulk solution and thus the close approach of the two surfaces results in the net displacement of solvent molecules from a region of higher chemical potential to a region of lower chemical potential (i.e., bulk solution); this is the origin of the attractive force. The equations for two planar surfaces or two spheres have been derived by Fleer et al.[15]

The corresponding expression for a sphere approaching a flat surface is readily derived, i.e.,

$$V_{dep}(h) = -P\left[\frac{2\pi}{3}(a + \Delta)^3 - (a + \Delta)^2 (a + h - \Delta) + \frac{\pi}{3}(a + h - \Delta)^3\right] \quad (19)$$

In Equation 19, P is the osmotic pressure of the bulk polymer solution; this may be determined experimentally or from existing theoretical expressions[16] and Δ is the thickness of the depletion layer. At low polymer concentrations (i.e., $C_2 < C_2^*$, the overlap concentration), $\Delta = r_g$, the radius of gyration of the polymer molecules concerned; for $C_2 > C_2^*$, Δ decreases in a complex manner with C_2 (the reader is referred to Reference 15 for details). With microorganisms one is frequently concerned with polyelectrolytes rather than neutral polymers. Basically speaking, similar effects (i.e., steric, bridging, depletion) occur, but these are more difficult to quantify, because of the complex nature of the electrostatic interactions involved. In general, however, all the effects referred to are enhanced; for example, bridging may be increased by strong Coulombic interactions between oppositely charged segments and surface groups or specifically interacting counter ions. Metal cations for example, have been shown to gel or precipitate a number of polysaccharides isolated from marine bacteria.[17]

In many cases microorganisms will come into contact with substrata (either animate or inanimate) coated with polymer layers which are able to interact specifically with those present on the surface of the microorganism. This may occur directly as in the case of receptor-adhesion interactions,[18] or via a mediating molecule, as postulated by Beachey et al.,[19] where deacylated lipoteichoic acid (LTA) is thought to bind two LTA binding proteins in the adsorption of Streptococci to animal cells.

C. Rate Constants

Both $\overrightarrow{k}$ and $\overleftarrow{k}$ depend on the nature of the overall microorganism/substratum interactions. We shall consider first the situation at low coverages, where lateral interactions between adsorbed microorganisms may be ignored.

1. Low Coverage

The simplest case to be considered is that where the microorganism "falls" into an infinite potential energy well at the surface (i.e., $V \rightarrow -\alpha$ at $h = a$, $V = 0$ for $h > a$, where h is the distance of the microorganism from the substratum surface), $\overleftarrow{k}$ is then zero (i.e., adsorption is irreversible). This is directly analogous to the condition considered by von Smoluchowski in his theory of particle coagulation.[5,20] J (flux per unit area of substratum surface) is then given by

$$J = D \cdot dC/dx \approx D \cdot C/a \tag{20}$$

where D is the diffusion coefficient of the particles; for dilute dispersions of noninteracting particles,

$$D = kT/f \tag{21}$$

where f is the friction coefficient of the particles:

$$f = 6\pi\eta\, a \tag{22}$$

where η is the viscosity of the medium. Hence,

$$J = kTC/6\pi\eta\, a^2 \tag{23}$$

Now, comparing Equations 4 and 5,

$$\overrightarrow{k}C = JS/n \tag{24}$$

S/n is the area per site, i.e., $\sim 4a^2$. Hence,

$$\overrightarrow{k} \sim 2kT/3\pi\eta \tag{25}$$

(i.e., $\overrightarrow{k} \sim 8.2 \times 10^{-19} m^3/sec$ for aqueous media at 20°C; see for example $\overrightarrow{k}$ for von Smoluchowski coagulation is $5.4 \times 10^{-18}\ m^3/sec$).

If the microorganism and substratum are charged and have the same sign, then one has to consider a combination of electrostatic repulsion and dispersion attraction. The resultant V(h) curve is shown in Figure 3A, for low electrolyte concentrations. There may be some (reversible) coagulation into the secondary minimum (V''_{min}). This case will be considered later. If $V''_{min} < \sim 1$ kT we may neglect this, and then aggregation into the primary minimum (V'_{min}) is the important consideration. Again, if this is sufficiently deep ($V''_{min} > \sim 10$ kT, say) then adsorption will be essentially irreversible (i.e., $\overleftarrow{k} = 0$, again); $\overrightarrow{k}$ is controlled by V_{max}.

By analogy with Fuchs'[21] analysis of the case of coagulation between charged particles with long-range interactions, the net flux, J, of charged microorganisms toward a charged substratum and also the microorganism concentration profile, C(h), may be calculated:

$$J = D \cdot dC/dh - C/f \cdot F(h) \tag{26}$$

where $F(h) = -dV/dh$, i.e., the force between the microorganisms and substratum.

Equation 26 is a first-order differential equation, for which a standard solution exists.[22] Making use of the boundary condition that for $h \rightarrow \alpha$, $C \rightarrow C^b$ (bulk concentration) and $V(h) \rightarrow 0$, then the following solution is obtained:

$$C = \frac{C^b}{\exp \frac{V(h)}{kT}} + \frac{J/D}{\exp \frac{V(h)}{kT}} \int_0^\infty \exp \frac{V(h)}{kT}\, dh \tag{27}$$

with $$J = \left(\frac{Dc^b}{a}\right) \Big/ \left[\frac{1}{a}\int_0^\infty \left(\exp \frac{V(h)}{kT}\right) dh\right] \tag{28}$$

or $$J = \frac{Dc^b}{aW} \tag{29}$$

where $$W = \frac{J_o}{J} = \frac{\overrightarrow{k}_o}{\overleftarrow{k}} = \frac{1}{a}\int_0^\infty \left(\exp \frac{V(h)}{kT}\right) dh \tag{30}$$

W may be identified as a "stability factor"; J_o and k_o are the microorganism flux and rate constant, respectively, for adsorption in the absence of long-range interaction forces.

If the substratum and microorganism are of opposite sign then V(h) is purely attractive, as illustrated in Figure 3B. Indeed, for low electrolyte concentrations $V_d(h)$ may be neglected in comparison to $V_e(h)$. As may be readily seen from Equation 26, the microorganism flux is now enhanced compared to the pure diffusion case, and $W < 1$.

When an adsorbed polymer layer is present on the microorganisms and/or substratum then, as we have seen, the substratum/microorganism interaction may be modified in various ways (depletion, bridging, etc.), but the more general form for V(h) is that depicted schematically in Figure 3C, i.e., a curve now characterized by a shallow minimum, V_{min}, such that adsorption is now, in general, reversible. (The situation is broadly similar for secondary minimum adsorption in the case of charge-stabilized systems: Figure 3A.) One now has to take account of $\overleftarrow{k}$ (see Equations 4 to 7).

The simplest interpretation of $\overleftarrow{k}$ is, perhaps, the one proposed by de Boer[23] in the context of molecular adsorption from the gas phase onto solid surfaces, i.e.,

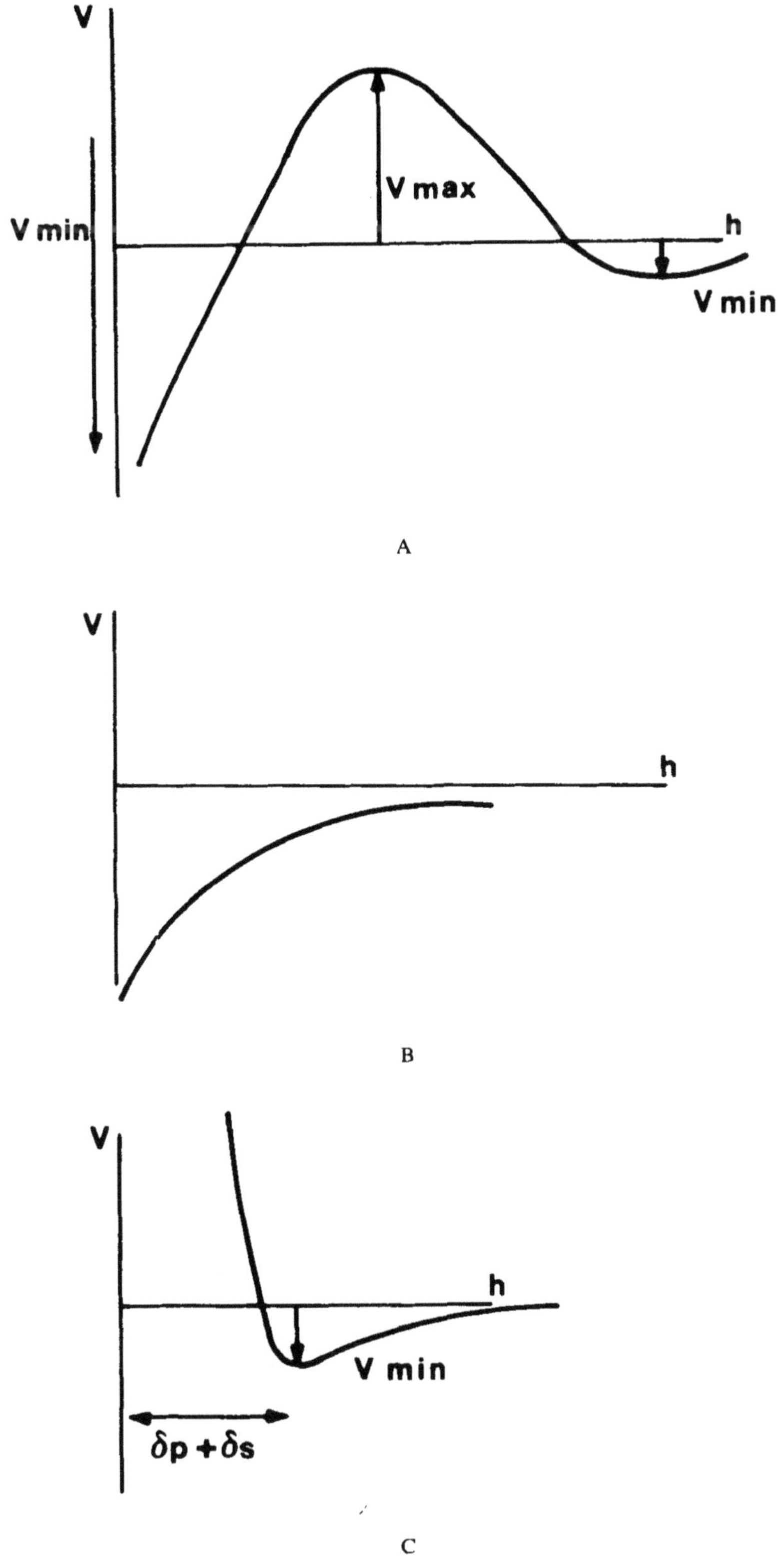

FIGURE 3. The variation of interaction energy V(h) with separation h.

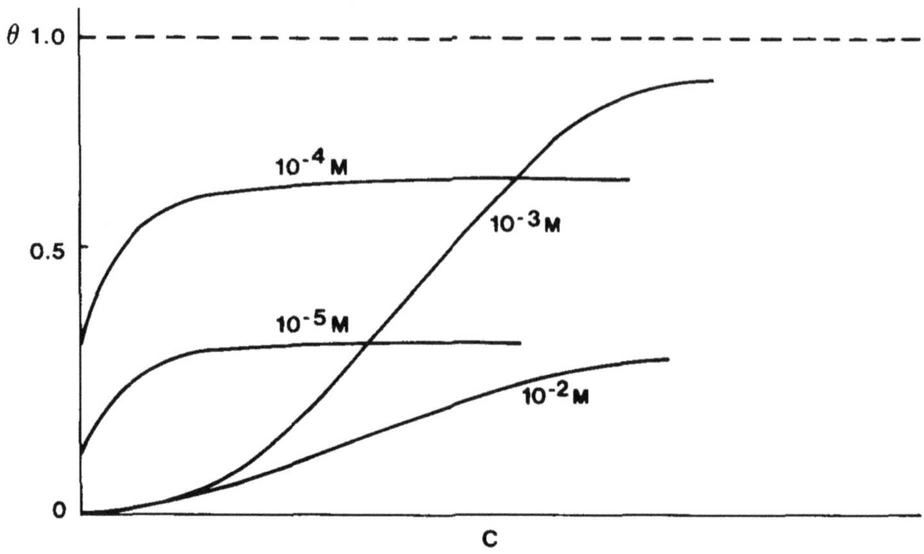

FIGURE 4. Equilibrium adsorption isotherms of small (0.2 μm diameter) positive latex particles onto larger (2.0 μm diameter) negative latex particles both possessing preadsorbed layers of poly(vinyl) alcohol at varying salt concentrations.

$$\overleftarrow{k} = \frac{1}{\tau} = \frac{1}{\tau_o} \cdot \exp\left(\frac{V_{min}}{kT}\right) \tag{31}$$

where τ is the surface residence time, and τ_o is the residence time in the absence of any interactions between the particles and the substratum. Unfortunately, it is not easy to ascribe a value to τ_o for particles on a surface, but Equation 31 does at least suggest a useful scaling relationship. A very rough estimate of the order of magnitude of τ_o could be made by considering the time it takes a particle to diffuse a certain fraction of its radius, fa,

$$\text{i.e., } \tau_o = \frac{(fa)^2}{D} = f^2\left[\frac{6\pi a^3\eta}{kT}\right] \tag{32}$$

For a 1-μm particle at 20°C in water, τ_o = 4.7 f^2 seconds. Hence, one might expect τ to be typically ~msec or sec for particles of this size.

2. High Coverage

At high coverage one has to take account of lateral interactions. Vincent et al.[24,25] showed that the form of the equilibrium particle adsorption isotherm for small positive polystyrene lattices onto much larger negative latex particles (in which both sets of particles carry preadsorbed layers of poly(vinyl)alcohol), depends very strongly on the electrolyte concentration (Figure 4). At low electrolyte concentrations, there is a lateral repulsive interaction between neighboring small particles adsorbed onto the surface of the large ones. This reduces the maximum attainable coverage to values as low as 0.2 (at 10^{-5} mol/dm^3 NaCl). Clearly, the calculation of V(h), and hence $\overrightarrow{k}$, for a particle approaching a negative surface in the situation where that surface already has many adsorbed particles present (Figure 5), is extremely complex. At large h values, one could use the net value of the zeta potential of the adsorbate surface (i.e., the large particles) for Ψ_s in the equations for V_e (Equations 13 or 14); but this approximation becomes increasingly invalid as h becomes smaller. Vincent

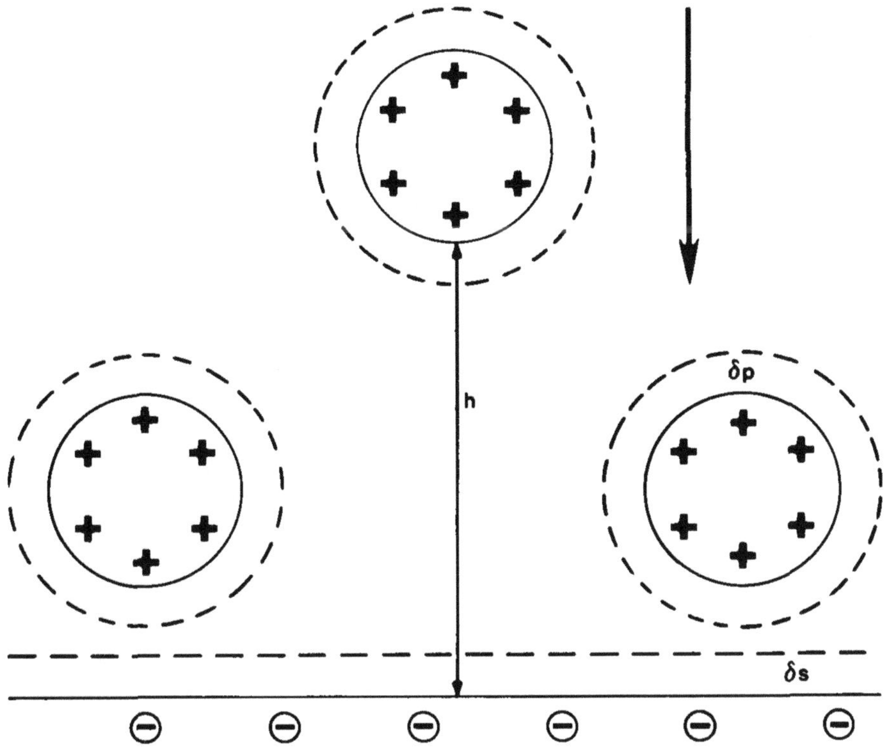

FIGURE 5. Lateral interactions between similarly charged particles carrying preadsorbed polymer layers.

et al.[24,25] observed that beyond a critical electrolyte concentration, the shape of the particle adsorption isotherm changed abruptly from the high affinity type to the low affinity type (Figure 4). This transition was identified with a change in the lateral interactions between neighboring adsorbed small particles from one of net repulsion to net attraction with increasing electrolyte concentration. Indeed, scanning electron microscopy[24] (SEM) revealed that at low electrolyte concentrations the adsorbed particles were well spaced, indicative of lateral repulsion, while at high electrolyte concentrations they formed two-dimensional clusters on the surface of the substratum, strongly suggestive of net lateral attraction (Figure 6).

Under conditions of net lateral attraction, where long-range electrostatic interactions are effectively suppressed, the calculation of V(h) would seem, in principle, to be more straightforward. To a reasonable approximation, one may simply add the substratum/particle and lateral particle/particle interactions together. Even so an attempt by Vincent et al.[25] to fit low affinity particle isotherms to a suitably modified Hill-de Boer equation[26] (developed for gas adsorption and based on a two-dimensional gas phase, with lateral interactions, model) was not very successful. Clearly, further work is necessary to fit particle adsorption (equilibrium) isotherms, at high coverages, through more accurate modeling of the overall interaction energies, before attempts can be made to predict $\overrightarrow{k}$ and $\overleftarrow{k}$ under these conditions.

D. Adsorption in Open Systems

Many natural systems are open and, hence, the number of microorganisms in the bulk fluid phase does not depend upon the concentration of free cells in the immediate vicinity of any substratum. This would be true of most river and marine environments, many industrial cooling systems, and adsorption from the air. Under these conditions the abstraction by

FIGURE 6. Small positive latex particles adsorbed to large negative latex particles from 10^{-4} *M* NaCl solution.

adsorption, or the reintrainment by desorption, of microorganisms does not significantly alter the bulk phase concentration C, i.e., C does not vary with time. Other factors, e.g., cell growth due to nutrient supply, could change C, but these are special cases and would have to be dealt with individually. The calculation of the deposition rate of the microorganisms $d\theta/dt$ is considerably simplified if C and, hence J are constant.

Recalling Equation 4,

$$\frac{d\theta}{dt} = \frac{JS(1 - \theta)}{n} - \overleftarrow{k}\,\theta$$

Integration leads to

$$\theta(t) = \left(\frac{JS/n}{JS/n + \overleftarrow{k}}\right) - \left(\frac{JS/n}{JS/n + \overleftarrow{k}}\right) \exp\left(-\frac{JS/n}{n} + \overleftarrow{k}\right) t \tag{33}$$

or if $\overleftarrow{k} = 0$ (irreversible adsorption),

$$\theta(t) = 1 - \exp\left(-\frac{JS}{n}\right) t \tag{34}$$

Again, therefore, the problem reduces to calculating J under various conditions. In quiescent environments microorganisms may sediment as well as diffuse toward horizontal substrata surfaces. The net flux will therefore be given by

$$J = J_S + J_D \tag{35}$$

where J_S is the contribution from sedimentation and J_D the contribution from diffusion. Since, in an open system, the concentration remains uniform up to the surface, $J_S = uC$ where u is the terminal velocity, given by Stokes' equation,

$$u = \frac{2a^2\Delta\varphi\ g}{9\eta} \tag{36}$$

where $\Delta\varphi$ is the density difference between the microorganism (or other particles) and the fluid medium, g is the acceleration due to gravity, and η is the fluid phase viscosity. J_D is again given by Equation 20. Hence,

$$J = C\left[\frac{2a^2\Delta\varphi\ g}{9\eta} + \frac{kT}{6\Pi\eta a^2}\right] \tag{37}$$

or for water at 298°K

$$J = C\left(2.45 \times 10^6\ a^2\Delta\varphi + \frac{2.45}{a^2} \times 10^{-19}\right) \tag{38}$$

Values of J_S, J_D, and J are plotted as a function of a for two values in Figure 7. As can be seen, and as might be expected, J_D dominates for small particles, while J_S dominates for large particles. For $\Delta\varphi$ ~0.1 kg/m^3 and a ~1 μm J_S and J_D are comparable in magnitude.

In the case of motile organisms a further term is required on the right side of Equation 35, to take account of the active movement of microorganisms. This is difficult to include in a quantitative manner, however.

In experiments where suspensions of a marine organism were allowed to sediment under gravity in a petri dish, Fletcher[27] showed that the number of cells which remain attached after rinsing is dependent upon C and also time, as expected. For $C = 3 \times 10^{15}/m^3$ Γ_{max} (reached after 2 hr) was found to be $4 \times 10^{11}/m^2$. Assuming each microorganism to be a sphere of radius 0.5 μm, θ_{sat} is approximately 0.3. Somewhat lower values (0.5 to $2 \times 10^{11}/m^2$) were obtained in experiments with plastic surfaces where these were mounted vertically, i.e., eliminating J_S.[28] Although there were a number of differences between the experiments, including the type of bacteria, these results seem to support the hypothesis that, in the case of bacteria, J_s and J_d are likely to be similar in magnitude as indicated by Equation 38. The maximum coverage (θ_{sat}) of 0.3 is illustrative of the common observation that it is unusual for more than 5 to 10% of the available substratum surface area to be covered with adsorbed cells (in the absence of surface growth). This may be a result of lateral interactions between adsorbing microorganisms (see Section II.B) or the simultaneous adsorption of cell surface components which progressively inhibit cell adsorption.

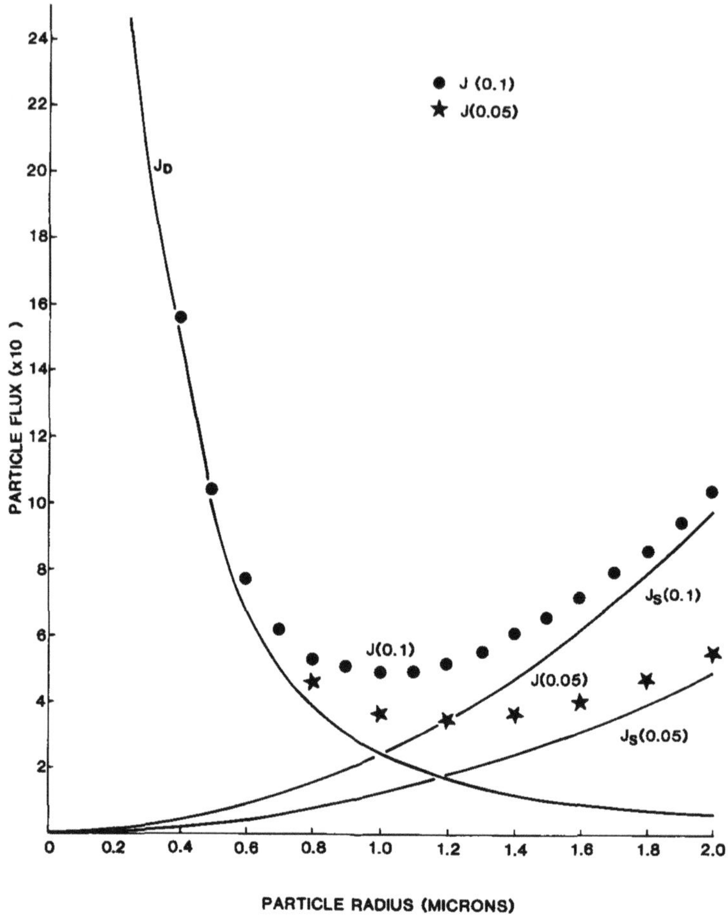

FIGURE 7. Variation in particle flux (J) due to sedimentation (J_S) and diffusion (J_D) with particle radius at density differences of 0.1 and 0.5 × 10^3/kg^3.

III. ADSORPTION UNDER FLUID FLOW CONDITIONS

A. Laminar Flow Conditions

Under flowing conditions the motion of the fluid over a collector surface serves to supply particles to the surface and also applies a shear stress to those already adsorbed. In laminar flow the flux of particles normal to the surface depends principally upon the flow rate, the geometry of the system, and the particle size. Various equations have been used to describe the accumulation of microorganisms on surfaces in various hydrodynamic conditions. These range from precise calculations of the particle flux to simple empirical expressions. In general terms the flux can be described by

$$J = BC \tag{39}$$

where B is some form of mass transfer coefficient. The derivation of B is complex and depends on the geometry of the systems under consideration. A popular experimental system for controlled deposition experiments consists of a planar disk-shaped collector mounted axially at the end of a shaft. As the disk is rotated in a suspension of particles a flux is set up which depends upon the speed of rotation according to the equation

$$J = 0.62\ D^{2/3}\ \nu^{-1/6}\omega^{1/2}C \tag{40}$$

where D is the particle diffusion coefficient, ν is the kinematic viscosity of the fluid, and ω is the angular velocity of the disk.

The rotating disk system has been used successfully to predict the rate of accumulation of particles on a surface in the absence of repulsive interactions.[30] It is less successful when the particles and substratum surface are similarly charged,[30] although the controlled flux still provides a useful experimental system for deposition studies and has been used in the examination of the adhesive behavior of microorganisms.[31,32] In these experiments an adsorbed amount of $5 \times 10^9/m^2$ was obtained after 20 min rotation at 160 rpm in a suspension of *Streptococcus mutans* containing $1.3 \times 10^{14}/m^3$. The Levich analysis assumes a stationary fluid phase; if in addition the fluid is moving, the particles may arrive at the surface as a result of interception, as well as by diffusion.[33] Interception also becomes important as the particles increase in size and become more easily influenced by fluid flow.

Dabros and van de Ven[34] have developed a radial flow chamber in which the suspension under test impinges on the center of a collector surface and then flows radially to the edge. Under these conditions the nondimensional (reduced) flux of particles (parallel to the surface) is

$$Sh = Ja/DC \tag{41}$$

where Sh is the Sherwood number, which can be expressed as:

$$Sh = \frac{\exp\left[-\frac{Pe}{6}(\delta + 1)^3 - Gr \cdot \delta\right]}{\int_\delta^\infty \exp\left[-\frac{Pe}{6}(\delta + 1)^3 - GrH\right] dH} \tag{42}$$

Pe, the Peclet number is $2\alpha\ a^3/D$ where α (the vorticity) is a constant, $H = (z - a)/a$ where z is the distance from the substratum surface and δ is usually identified with the distance of the primary energy minimum from that surface (see Figure 3).

Gr is a dimensionless gravity number, defined by

$$Gr = \frac{2}{9}\frac{\Delta\varphi\ ga^3}{\eta D} \tag{43}$$

This equation requires considerable modification to take account of particle desorption and blocking effects before reasonable agreement with experimental deposition rates is obtained, even in the case of well-characterized, nonbiological particles. Other expressions for the flux of particles under various geometries, e.g., parallel channel, plate in uniform flow, have been reviewed by Adamczk et al.[1]

The possible removal of particles under shear forces generated by the action of the fluid flowing parallel to the surface must also be considered and has been the subject of a considerable amount of experimental work. Much of this has been directed to identifying flow regimes which might prevent or reduce the buildup of microbial deposits in pipe flow systems such as heat exchangers.

A particle adsorbed to a surface over which a fluid flows experiences a drag force, F_h,[35] where

$$F_h = 1.7005\ \pi\eta au \tag{44}$$

Table 1
THE EFFECT OF WALL SHEAR STRESS ON THE ACCUMULATION OF MICROORGANISMS ON SURFACES

Organism	Surface	Wall shear stress required to remove attached cells (N/m^2)	Wall shear stress required to prevent cell attachment (N/m^2)	Ref.
Chlorella	Glass	0.081	ND	38
	Glass (in 2×10^{-5} *M* $FeCl_3$)	81	ND	
Bacillus cereus	Glass	ND	28	41
	Siliconized glass	ND	52	
Streptococcus sanguis	Glass	ND	28	43
Gram + ve cocci	Glass (Pyrex®)	45	ND	45
	Siliconized glass	26	ND	
	Steel	54	ND	
Pseudomonas fluorescens	Stainless steel	12	6—8	39

ND = no data available.

Under conditions of laminar flow,

$$u = (T/\eta)\,a \tag{45}$$

where T is the wall shear stress.

In flow through a cylindrical pipe,

$$T = -(du/dy) \tag{46}$$

$$= 4.Q\eta/\pi a^3 \tag{47}$$

where Q is the volumetric flow rate.[36]

The drag force operates parallel to the surface and is often equated to the force of adhesion F_a. There seems little justification for this, however, since under laminar flow conditions there is no force component directed away from the surface.[37] Visser[35] dealt with this difficulty by comparing the removal of carbon particles from a cellophane surface by shear forces with the removal under normal forces generated in a centrifuge. He concluded that as normal forces F_c and hydrodynamic forces F_h of similar magnitude removed a similar percentage of particles, both could be equated to the force of adhesion F_a.

Prior to Visser's work, Nordin and co-workers[38] had carried out experiments on the adsorption of *Chlorella* to the walls of a parallel-sided glass tube. They showed that cells which had been allowed to settle onto the glass wall of the tube and remain there for 2 hr could be removed by passing fluid through the tube at average flow rates of above 10^{-8} m^3/sec. The wall shear stress required to move the cells was strongly dependent on the ionic strength and composition of the fluid medium. Under conditions where the glass and *Chlorella* surfaces were oppositely charged a 100-fold increase in flow rate was required to detach the cells (see Table 1). At these high flow rates the flow was almost certainly turbulent.

Other workers using various devices have attempted to measure both the force required to remove attached microorganisms and that required to reduce the rate of adsorption to zero. Duddridge et al.[39] demonstrated experimentally that a lower shear stress is required to prevent adsorption than to remove microorganisms which have been in contact with the surface for some time (see Table 1). This supports the idea that changes occur at the cell/substratum junction which gradually consolidate the adhesive joint. The initial interaction between a microorganism and collector will be influenced by the interplay of electrostatic, dispersion, and structural forces, as described in Section II.B. During the early stages of adsorption, cells are often seen to oscillate over a small area of the collector surface. This is unlikely to be secondary minimum capture since in at least one example the adsorption was insensitive to changes in ionic strength.[40] In addition, cells moving across a surface under the influence of a flowing suspension often appear to pause for a few seconds before moving on.[41,42] Powell and Slater[41] attribute this to the formation of weak polymer linkages with the surface which are unable to fully resist the force of the fluid. This supports the work of Rutter and Leach,[43] which suggested that a certain residence time at a point on a collector surface was required for the establishment of sufficient polymer bridges to anchor the cell in position.

"Reversible" and "irreversible" adsorption may therefore simply reflect different stages of the same time-dependent adsorption process. They appear to be different because in order to determine the number of adsorbed cells a shear force is applied which divides the organisms into two groups; those which detach and those which do not. The relative proportions of the cells on a surface which fall into the two groups will depend upon the time they have spent in contact with the surface. The time required to achieve a strong interaction with the surface will probably depend upon the nature and concentration of cell surface polymers capable of adsorbing to the collector surface. Other changes may also occur due to the proximity of the surface. These may be active, involving cell metabolism, or passive, involving reorganizations in the adsorbing polymers. Surprisingly, perhaps, Duddridge et al.[39] showed that the wall shear stress required to remove attached cells from stainless steel was only twice that required to prevent attachment (see Table 1). Unfortunately there are few data available concerning the change in the strength of microbial adhesion with time. It is difficult, therefore, to speculate on the nature of any consolidation processes which may occur or at which point an adhesive joint fractures. This would be a fruitful area for further study.

Under flowing conditions the initial rate of removal of particles is a function of t, thus $\overleftarrow{k}$ will be related to t. Unfortunately, since the adhesive interactions between the microorganisms and the substratum surface appear, in most cases, also to increase with time, $\overleftarrow{k}$ will be a complex function depending upon surface coverage, the wall stress T, and the cell residence time. Equations describing the rate of accumulation of microorganisms on surfaces in practical situations will therefore usually be empirical with regression coefficients drawn from experience rather than theory.

In an open flowing system the rate at which microorganisms accumulate on a substratum surface involves a large number of interacting processes. These may be grouped as follows to give a qualitative description of the way in which a microbial deposit builds up.

$$dN/dt = \text{adsorption rate} - \text{desorption rate} + \text{surface growth rate} \tag{48}$$

Recalling Equation 4

$$dN/dt = J\,S(N_{max} - N)/n - k(T)\,N + \mu sN \tag{49}$$

where N is the number of microorganisms on the surface at time t, N_{max} is the number at saturation coverage, and μs is the specific surface growth rate.

In a flowing system, k depends upon the strength of interaction with the substratum surface, which may vary with time, and also the wall shear stress.

Equation 49 can be made to fit the observed asymptotic adsorption kinetics of microorganisms by inserting suitable values for J and k. However, this approach may not yield any information regarding the actual processes of consolidation occurring at the microorganism/substratum interface or the reasons for surface saturation. Surface saturation may, for example, be the result of lateral interactions due to adsorbed microorganisms rather than the establishment of equilibrium between adsorbing and desorbing cells.

B. Turbulent Flow Conditions

In many environments where the adsorption of microorganisms can cause serious problems in terms of loss of efficiency, corrosion, etc. (e.g., heat exchangers) turbulent flow occurs rather than laminar flow. Under turbulent flow the thickness of the liquid boundary layer adjacent to the pipe wall is much thinner than under laminar flow conditions. This results in a much steeper velocity gradient across the boundary film leading to an increase in the deposition rate.

The deposition and removal of small particles in turbulent flow have received considerable attention from engineers, particularly with regard to the deposition of particles under the influence of inertial and impact mechanisms. Visual observation of particles in the proximity of tube walls in turbulent flow regimes shows that some particles suddenly move almost at right angles away from the surface and into the flowing stream. This behavior led Cleaver and Yates[44] to suggest that the viscous sublayer is continuously disrupted by turbulent "bursts". Consideration of the inertial motion and diffusion of small particles under these conditions led to the following quantitative expression for the total particle flux,

$$J = J_i + J_d \tag{50}$$

or

$$J = \frac{a^2U^{*2}}{11.8\,\nu^2} + \frac{0.014kT}{\pi\varphi\nu^2 a} \tag{51}$$

where U* is the wall friction velocity $((\tau/\varphi)^{\frac{1}{2}})$ and ν is the kinematic viscosity (η/φ).

An important aspect of Cleaver and Yates' theory is that the turbulent bursts can generate a lift force normal to the surface sufficient to detach adsorbed particles. The normal lift force acting on a particle is given by:

$$F_l = 0.608\varphi\,\nu^2\left(\frac{aU^*}{\nu}\right)^3 \tag{52}$$

For small particles, $U^*/\nu < 1$, the lift force F_l is less than the drag force F_h.[37] It is possible, therefore, that the adhesive link between the particle and the substrate is initially broken by drag forces before the particle is transported away from the proximity of the surface by a turbulent burst. In both cases, however, the force acting to remove the adsorbed particles is a function of the wall shear stress.

By considering the frequency and spatial arrangement of the turbulent bursts they were able to calculate the rate at which particles would be removed from a surface. This analysis is rather complex but shows that the rate of removal of particles from the surface in turbulent flow is, like laminar flow, proportional to the wall shear stress.

ACKNOWLEDGMENT

P. Rutter wishes to thank the British Petroleum Company for permission to publish this material.

REFERENCES

1. **Adamczyk, Z., Dabros, T., Czarnecki, J., and Van de Ven, T. G. M.,** Particle transfer to solid surfaces, *Adv. Colloid Interface Sci.*, 19, 183, 1983.
2. **Langmuir, I.,** Adsorption of gases onto plane surfaces of glass, mica and platinum, *J. Am. Chem. Soc.*, 40, 1361, 1918.
3. **Boughley, M. T., Duckworth, M. R., Lips, A., and Smith, A. L.,** Observation of weak primary minima in the interaction of polystyrene particles with nylon fibres, *J. Chem. Soc. Faraday Trans. 1*, 74, 2200, 1978.
4. **Vincent, B., Jafelicci, M., Luckham, P. F., and Tadros, Th. F.,** Adsorption of small positive particles onto large negative particles in the presence of polymer. II. Adsorption equilibrium and kinetics as a function of temperature, *J. Chem. Soc. Faraday Trans. 1*, 76, 674, 1980.
5. **Overbeek, J. Th.,** in *Colloid Science*, Vol. 1, Kruyt, H. R., Ed., Elsevier, Amsterdam, 1952, 278.
6. See Reference 5, p. 264.
7. **Hogg, R., Healy, T. W., and Fuerstenau, D. W.,** Mutual coagulation of colloidal dispersions, *Trans. Faraday Soc.*, 62, 1638, 1966.
8. **Vincent, B. and Whittington, S.,** in *Surface and Colloid Science*, Vol. 12, Matijevic, E., Ed., Plenum Press, New York, 1982, 1.
9. **Napper, D. H.,** *Polymeric Stabilisation of Colloidal Dispersions*, Academic Press, New York, 1983.
10. **Scheutjens, J. M. H. M. and Fleer, E. J.,** in *The Effects of Polymers on Dispersion Properties*, Tadros, Th. F., Ed., Academic Press, New York, 1982, 145.
11. **Rutter, P. R.,** The physical chemistry of the adhesion of bacteria and other cells, in *Cell Adhesion and Motility*, Curtis, A. S. G. and Pitts, J. D., Eds., Cambridge University Press, London, 1980, 103.
12. **Rutter, P. R. and Vincent, B.,** Physicochemical interactions of the substratum, microorganisms, and the fluid phase, in *Report of the Dahlem Workshop on Microbial Adhesion and Aggregation*, Marshall, K. C., Ed., Springer-Verlag, Berlin, 1984, 21.
13. **Fletcher, M. and Floodgate, G. D.,** An electron-microscope demonstration of an acidic polysaccharide involved in the adhesion of a marine bacterium to solid surfaces, *J. Gen. Microbiol.*, 74, 325, 1973.
14. See Reference 8, p. 72 or Reference 9, p. 332.
15. **Fleer, G. J., Scheutjens, J. M. H. M., and Vincent, B.,** in *Polymer Adsorption and Dispersion Stability*, Goddard, E. D. and Vincent, B., Eds., American Chemical Society, Washington, D. C., 1984, 245.
16. **Flory, P. J.,** *Principles of Polymer Chemistry*, Cornell University Press, Ithaca, N.Y., 1953, 495.
17. **Sutherland, I. W.,** Polysaccharides in the adhesion of marine and freshwater bacteria, in *Microbial Adhesion to Surfaces*, Berkeley, R. C. W., Lynch, J. M., Melling, J., Rutter, P. R., and Vincent, B., Eds., Ellis Horwood, Chichester, England, 1980, 329.
18. **Jones, G. W.,** Adhesion to animal surfaces, in *Report of the Dahlem Workshop on Microbial Adhesion and Aggregation*, Marshall, K. C., Ed., Springer-Verlag, Berlin, 1984, 7.
19. **Beachey, E. M., Simpson, W. A., and Ofek, I.,** Interaction of surface polymers of *Streptococcus pyogenes* with animal cells, in *Microbial Adhesion to Surfaces*, Berkeley, R. C. W., Lynch, J. M., Melling, J., Rutter, P. R., and Vincent, B., Ellis Horwood, Chichester, England, 1980, 389.
20. **Von Smoluchowski, M.,** Versuch einer mathematischen Theorie der Koagulationskinetik Kolloidaler Lösungen, *Z. Physik. Chem.*, 92, 129, 1917.
21. **Fuchs, N.,** Über die Stabilität und Aufladung der Aerosole, *Z. Physik.*, 89, 736, 1934.
22. **Boas, M. L.,** *Mathematical Methods in the Physical Sciences*, Wiley Interscience, New York, 1966, 329.
23. **de Boer, J. N.,** *The Dynamic Character of Adsorption*, Clarendon Press, Oxford, 1953.
24. **Vincent, B., Young, C. A., and Tadros, Th. F.,** Equilibrium aspects of heteroflocculation in mixed sterically stabilised dispersions, *Faraday Disc. Chem. Soc.*, 65, 296, 1978.
25. **Vincent, B., Young, C. A., and Tadros, Th. F.,** Adsorption of small, positive particles onto large, negative particles in the presence of polymer. I. Adsorption isotherms, *J. Chem. Soc. Faraday Trans. 1*, 76, 665, 1980.
26. **Hill, T. L.,** Statistical mechanics of multi-molecular adsorption, *J. Chem. Phys.*, 14, 441, 1946.
27. **Fletcher, M.,** The attachment of bacteria to surfaces in aquatic environments, in *Adhesion of Microorganisms to Surfaces*, Ellwood, D. C., Melling, J., and Rutter, P. R., Eds., Academic Press, New York, 1979, 87.
28. **Pringle, J. M. and Fletcher, M.,** Influence of substratum wettability on attachment of freshwater bacteria to solid surfaces, *Appl. Environ. Microbiol.*, 45(3), 811, 1983.
29. **Levich, V. G.,** *Physicochemical Hydrodynamics*, Prentice-Hall, Englewood Cliffs, N.J., 1962.
30. **Hull, M. and Kitchener, J. A.,** Interaction of spherical colloidal particles with planar surfaces, *Trans. Faraday Soc.*, 65, 3093, 1969.
31. **Rutter, P. R. and Abbott, A.,** A study of the interaction between oral *Streptococci* and hard surfaces, *J. Gen. Microbiol.*, 105, 219, 1978.
32. **Abbott, A.,** Deposition of *Streptococcus mutans*, Ph.D. thesis, University of Bristol, Bristol, England, 1981.

33. **Spielman, L. A. and Cukor, P. M.,** Deposition of non-Brownian particles under colloidal forces, *J. Colloid Interface Sci.*, 43, 51, 1973.
34. **Dabros, T. and van de Ven, T. G. M.,** A direct method for studying particle deposition onto solid surfaces, *Colloid Polym. Sci.*, 261, 694, 1983.
35. **Visser, J.,** Measurement of the force of adhesion between submicron carbon-black particles and a cellulose film in aqueous solution, *J. Colloid Interface Sci.*, 34(1), 26, 1970.
36. **Schlicting, H.,** *Boundary Layer Theory*, 4th ed., McGraw-Hill, New York, 1960.
37. **Cleaver, J. W. and Yates, B.,** Mechanism of detachment of colloidal particles from a flat substrate in a turbulent flow, *J. Colloid Interface Sci.*, 44(3), 464, 1973.
38. **Nordin, J. S., Tsuchiya, H. M., and Fredrickson, A. G.,** Interfacial phenomena governing adhesion of *Chlorella* to glass surfaces, *Biotechnol. Bioeng.*, 9, 543, 1967.
39. **Duddridge, J. E., Kent, C. A., and Laws, J. F.,** Effect of surface shear stress on the attachment of *Pseudomonas fluorescens* to stainless steel under defined flow conditions, *Biotechnol. Bioeng.*, 24, 153, 1982.
40. **Abbott, A., Rutter, P. R., and Berkeley, R. C. W.,** The influence of ionic strength, pH and a protein layer on the interactions between *Streptococcus mutans* and glass surfaces, *J. Gen. Microbiol.*, 129, 439, 1983.
41. **Powell, M. S. and Slater, N. K. H.,** The deposition of bacterial cells from laminar flows onto solid surfaces, *Biotechnol. Bioeng.*, 15, 891, 1983.
42. **Doroszewski, J.,** Short term and incomplete cell-substrate adhesion, in *Cell Adhesion and Motility*, Curtis, A. S. G. and Pitts, J. D., Eds., Cambridge University Press, London, 1980, 171.
43. **Rutter, P. R. and Leech, R.,** The deposition of *Streptococcus sanguis* NCTC 7868 from a flowing suspension, *J. Gen. Microbiol.*, 120, 301, 1980.
44. **Cleaver, J. W. and Yates, B.,** A sublayer model for the deposition of particles from tubulent flow, *Chem. Eng. Sci.*, 30, 983, 1975.
45. **Fowler, H. W. and McKay, A. J.,** The measurement of microbial adhesion, in *Microbial Adhesion to Surfaces*, Berkeley, R. C. W., Lynch, J. M., Melling, J., Rutter, P. R., and Vincent, B., Eds., Ellis Horwood, Chichester, England, 1980, 143.

Chapter 11

MODELING BIOFILM ACCUMULATION

James D. Bryers

TABLE OF CONTENTS

I. INTRODUCTION

A. Biofilms in Natural and Engineered Systems

What are biofilms and are they significant? Answering such questions is reminiscent of seven blind Indians trying to describe an elephant, each touching only one part of the beast. No absolute definition of a biofilm exists, for a single rule would only be system-specific, thus creating numerous exceptions. A consensus at a recent Dahlem conference[1] on microbial adhesion and aggregation stated "a biofilm" is, in general, a collection of microorganisms and extracellular products associated with a solid (living or inanimate) surface, termed a substratum.

Consequently, biofilms can be either isolated cells or colonies in a sparse (less than 1 to 3%) coverage of soil detritus (Figure 1); a gelatinous polymer matrix, 10 μm to several hundred micrometers thick on a heat exchanger condenser tube (Figure 2); a bacterial-algal slime of several millimeters covering a rock submerged in an alpine stream (Figure 3); from bacterial coverage of the digestive tract (Figure 4); to dense slime layers on the internal surfaces of a fermenter (Figure 5). Essentially, any interface that exhibits microbial activity can be conceptually termed a biofilm.

Thus, biofilms can be found in any natural or man-made system exposed to an unsterile aqueous environment. It may well be that the majority of microbial processes in nature are mediated by microorganisms associated with some kind of interface. In engineered systems, experience has shown biofilms to be significant in either a beneficial or detrimental fashion. Characklis[5] provides an elegant summary of the effects and relevance of biofilms which is reiterated here in Table 1.

As one can see, biofilms affect any system, positively or adversely, either by their physical presence or through the microbial conversions they mediate. Therefore, any modeling of biofilm formation should consider: (1) development of the cells and the biofilm structure; (2) the change in surface metabolic activity; and, if pertinent, (3) the relationship between changing biofilm amount and an observed system response (e.g., pressure drop, corrosion rate, overall substrate removal rate).

B. Biofilm Development: A Conceptual Scenario

Experience suggests biofilms develop according to the scenario depicted in Figure 6.

A biofilm community is initiated when a "clean" surface is exposed to an aqueous environment and becomes conditioned by chemical constituents therein. Inevitably, microorganisms become associated with the surface, adhere, then attach tenaciously. Once these primary colonizers are firmly bound, the activity of the community is dependent on the metabolism and growth under local surface conditions. Such metabolic activities include substrate consumption, cellular replication, and synthesis of exopolymers. Thus, the biofilm gel matrix accumulates on the surface acting to trap elusive nutrients and, if in a mixed culture system, attract other microbial participants to the biofilm community. Successive microcommunities develop environmental niches within certain layers as the biofilm thickness increases. Eventually, the biofilm thickness reaches a steady state where processes producing more film are counterbalanced by processes reducing or removing biofilm. Continued conversion of substrate produces excess biomass and soluble products that are entrained into the surrounding fluid, perhaps acting to attract higher life forms, thus initiating further ecological succession.

This chapter (1) reviews the various processes contributing to overall biofilm development, (2) presents current mathematical descriptions of these individual processes, and (3) illustrates several mathematical simulations of biofilm development.

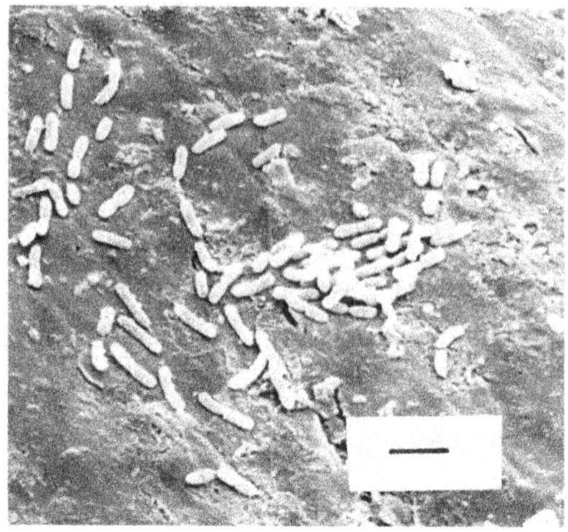

FIGURE 1. Sparse bacterial coverage of soil detritus recovered from groundwater aquifer.[2] Bar represents 3.0 μm.

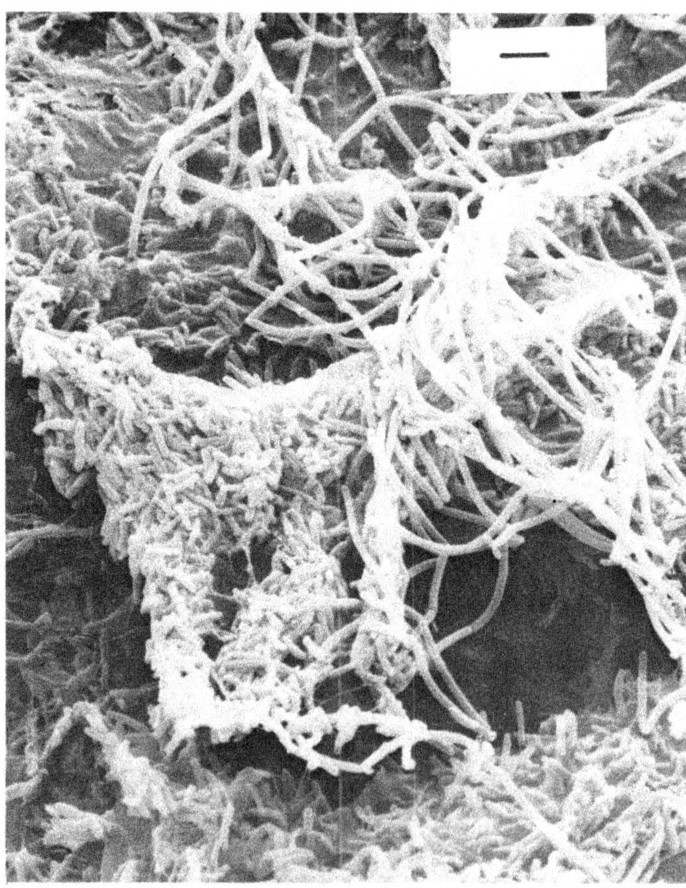

FIGURE 2. Bacterial biofilm coverage of a laboratory heat exchanger condenser tube. Biofilm measured 150 μm prior to SEM preparation. Bar represents 2.0 μm.

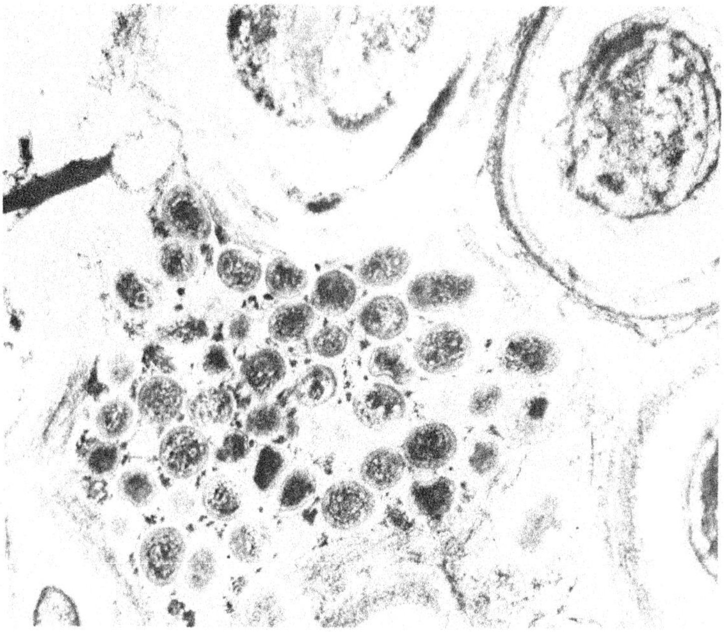

FIGURE 3. Electron micrograph of a slime-enclosed microcolony of aquatic bacteria which lies between the well-developed slime sheaths of several blue-green bacteria in an adherent matrix in an alpine stream. (From Geesey, G. G. et al., *Can. J. Microbiol.*, 23, 1733, 1977. With permission.)

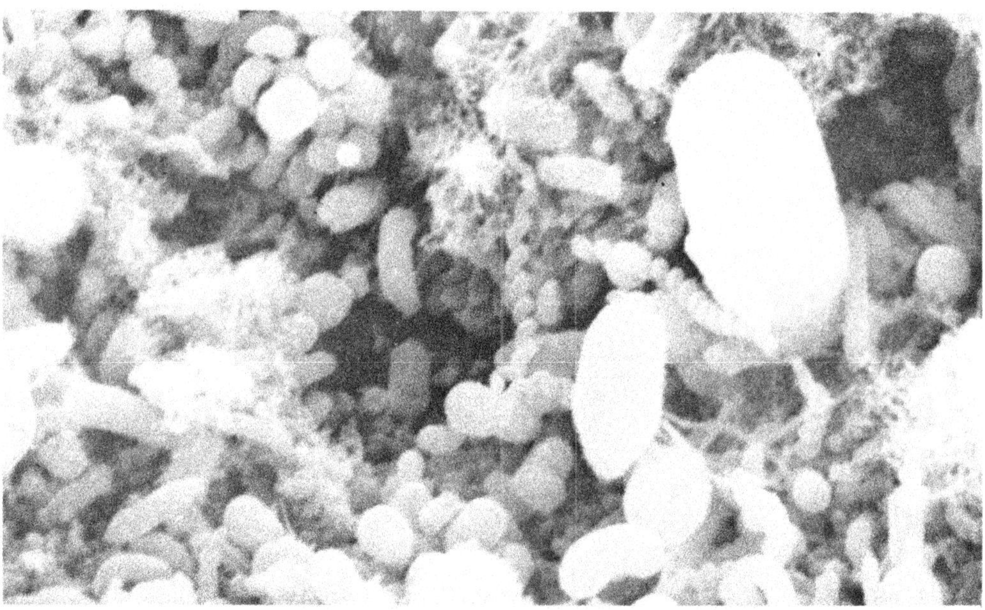

FIGURE 4. Mixed population of bacteria attached to the surface cells of the rumen of a cow. (From Costerton, J. W. et al., *Sci. Am.*, 238, 86, 1978. With permission.)

FIGURE 5. A pure culture biofilm of *Hyphomicrobium* spp. formed on the internal surfaces of a continuous culture bioreactor.

II. PROCESSES GOVERNING BIOFILM FORMATION AND PERSISTENCE

This section considers the various processes that are known to contribute to the overall formation and persistence of a biofilm, with special emphasis on the mathematical description of such processes.

The literature suggests biofilm formation is the net result of the following processes:

- Deposition-related processes
  - Surface preconditioning
  - Cellular particulate transport
  - Adhesion
- Metabolic processes
  - Substrate consumption, cellular growth, and Cellular replication
  - Maintenance
  - Extracellular polymer production
- Removal processes
  - Continual biofilm surface entrainment
  - Sloughing

There are several excellent review articles detailing these processes, that are more exhaustive than what will be presented in this section; the reader is urged to consult these articles for further detail.[1,5-11]

A. Deposition-Related Processes

Deposition is also a "net event", itself comprised of several individual processes: (1)

Table 1
EFFECT AND RELEVANCE OF BIOFILMS ON VARIOUS RATE PROCESSES[5]

Effects	Specific process and result	Concerns
Heat transfer reduction	Biofilm formation on condenser tubes and cooling tower fill material, *energy losses*	Power industry, chemical process industry, U.S. Navy, solar energy systems
Increase in fluid frictional resistance	Biofilm formation in water and wastewater conduits as well as condenser and heat exchange tubes; causes increased power consumption for pumped systems or reduced capacity in gravity systems, *energy losses*	Municipal utilities, power industry, chemical process industry, solar energy systems
	Biofilm formation on ship hulls causing increased fuel consumption, *energy losses*	U.S. Navy, shipping industry
Mass transfer and chemical transformations	Accelerated corrosion due to processes in the lower layers of the biofilm; results in *material deterioration* in metal condenser tubes, sewage conduits, and cooling tower fill	Power industry, U.S. Navy, municipal utilities, chemical process industry
	Biofilm formation on remote sensors, submarine periscopes, sight glasses, etc., causing *reduced effectiveness*	U.S. Navy, water quality quality data collection
	Detachment of microorganisms from biofilms in cooling towers Releases *pathogenic organisms* (e.g., *Legionella* in aerosols)	Public health
	Biofilm formation and detachment in drinking water distribution systems; changes *water quality* in distribution system	Municipal utilities, public health
	Biofilm formation on teeth, causes *dental plaque and caries*	Dental health
	Attachment of microbial cells to animal tissue, causes *disease* of lungs, intestinal tract, and urinary tract	Human health
	Extraction and oxidation of organic and inorganic compounds from water and wastewater (e.g., rotating biological contacters, biologically aided carbon adsorption and benthal stream activity), *reduced pollutant load*	Wastewater treatment, water treatment, stream analysis
	Biofilm formation in industrial production processes *reduces product quality*	Pulp and paper industry
	Immobilized organisms or community of organisms for conducting *specific chemical transformations*	Chemical process industry

Table 1 (continued)
EFFECT AND RELEVANCE OF BIOFILMS ON VARIOUS RATE PROCESSES[5]

Effects	Specific process and result	Concerns
	Fouling biofilm accumulation *reduces effectiveness* of ion exchange and membrane processes used for high quality water treatment	Desalination, industrial water treatment

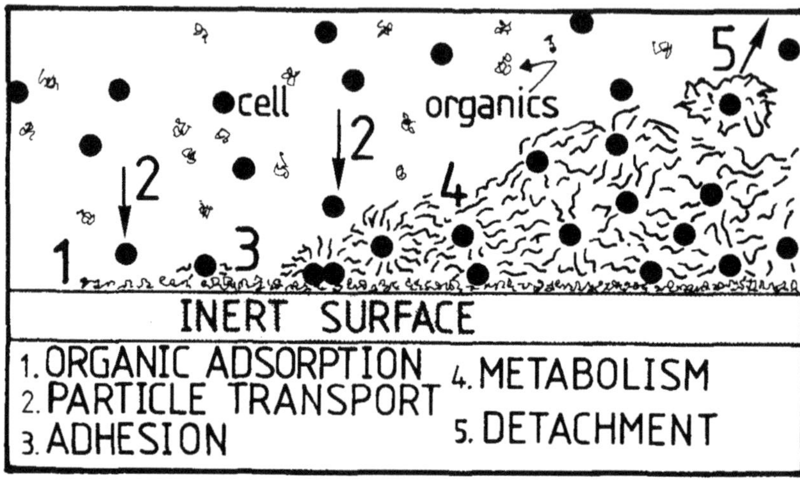

FIGURE 6. Processes contributing to biofilm formation and persistence.

macromolecular organic preconditioning of the virgin surface, (2) cellular transport from the bulk phase to the solid substratum, and (3) adhesion to the surface. Since most studies are only able to measure the net amount of cells remaining attached to a study surface, it is most unlikely that any of the above processes have been studied directly. Under the guise of quantifying "cellular adhesion" or "particle transport", results of most studies reflect only overall observed deposition rates. In the interim between preparation and the printing of this manuscript, an elegant study was completed by Escher,[101] that quantifies the kinetics of cell-particle transport and cell adhesion, separately, and not as a "lumped process".

1. Surface Preconditioning

Within minutes, perhaps seconds of exposure, adsorption of measurable quantities of organic molecules occurs. The total amount of the organics absorbed is insignificant in terms of a mature biofilm but this initial organic layer is sufficient to precondition the substratum, influencing further events. Such organic polymers have as many as 10^5 subunits; thus, in principle any one molecule could have that number of bonds with the substratum. Most likely, the number of polymer subunits associated with the surface is between 30 to 60% of the total chain length. Consequently, once a polymer is adsorbed to a surface, it is unlikely that all these bonds could be broken simultaneously without forming new ones, which negates polymer desorption.

These conditioning organics appear to be very specific macromolecules of microbial origin. Loeb and Neihoff[12] investigated adsorption of soluble components from seawater and its effect on the wettability and charge characteristics of several solid materials. Changes in surfaces exposed to seawater (with and without naturally occurring organics) were observed by microelectrophoresis, ellipsometry, and contact angle measurements. Ellipsometry results showed a rapid surface adsorption of a 0.03 to 0.08 μm organic film after 200 min exposure to natural seawater. (Film thickness estimates were dependent upon the value of refractive index assumed). Contact angles for platinum plates changed from clean values of less than 10° to values greater than 35° for seawater-exposed samples. Various types of solid particles, exhibiting a variety of electrophoretic mobilities (both positive and negative), all changed to a more restricted, moderately negative range upon exposure to natural seawater. Loeb and Neihoff concluded that rapid adsorption of naturally occurring organics is sufficient to modify the "attractiveness" of inert surfaces to attaching organisms.

Baier, in a series of papers, confirmed the development of an adsorbed organic film on inert surfaces prior to cell adhesion. Baier et al.[13] reviewed the surface chemistry and physics of bioadhesion emphasizing surface wettability. Baier and Weiss[14] demonstrated the involvement of surface adsorbed glycoproteins in tumor cell adhesion. Baier and Depalma[15] verified correlations between apparent surface-free energies of various materials and their degree of associated adsorbed biomass. Using human red blood cells, Baier[16] found a range of surface tensions, 25 to 30 dyn/cm, that enhanced cellular attachment. Application of a glycoprotein material to germanium test surfaces reduced the surface free energy to 30 to 40 dyn/cm, well within the range favoring biological adhesion.[14,15] Ellipsometry measurements indicate rapid saturation of the germanium surfaces by organic films approximately 0.006 μm thick.

Dexter and co-workers[17-19] provide data for a variety of solids exposed to warm coastal seawater for periods ranging from 20 to 384 hr. At each exposure time the reported number of bacteria was shown as a function of the existing critical surface tension. This work found the maximum cell count on high-energy surfaces and the minimum on low-energy surfaces. Dexter's work, as with Baier's, concludes that adsorbed organic layers shift initial values of critical surface tension toward a narrow range; the surface tension range compatible for bioadhesion appears unaffected by organic adsorption.

Sufficient evidence exists to indicate that the organic material "biasing" solid surfaces to attachment in natural environments is microbial in origin. Fletcher and Loeb[20] considered the effects of six different proteins on bacterial attachment to polystyrene petri dishes. Bovine serum albumin, gelatin, fibrinogen, and pepsin impaired the attachment of a marine pseudomonad, apparently by preconditioning only the polystyrene surface. Serum albumin, however, also appeared to modify the bacterial surface. Basic proteins, protamine, and histone did not markedly affect attachment. The inhibition of cellular adhesion by an adsorbed protein layer has also been reported by Valentine and Allison.[21] Contradicting Dexter's work, that of Fletcher and Loeb[20] found one marine pseudomonad NCMB 2021, attached preferentially to low-energy substratum.

Corpe[22] isolated the exopolysaccharide layer from an attached marine bacterium, *Pseudomonas atlantica,* coated glass slides on one side with 1 mg/mℓ of the polysaccharide solution and allowed them to air dry. Clean and polymer pretreated slides were then exposed to seawater resulting in an enhanced bacterial attachment on the pretreated surfaces. Other slides were smeared centrally with the polymer solution, air dried, and exposed to stagnant marine aquaria conditions. Observations indicated a distinct preferential attachment of marine bacteria on the portion of surface treated by the polymer.

Tosteson and Corpe[23] observed enhanced adhesion of washed *Chlorella vulgaris* cells to solid surfaces pretreated with adsorbed extracellular material recovered from either *Chlorella,* a marine bacterial culture, natural seawater, or fouled marine surfaces. Extracellular material isolated from bacterial cultures and marine fouled surfaces enhanced *Chlorella* adhesion by three orders of magnitude more than *Chlorella*-derived material. It should be noted that the extraneous materials were uniformly mixed with washed cells and not applied directly to test surfaces. Thus, results may reflect polymer influence on flocculation/sedimentation and cell surface characteristics rather than an enhancement of surface adhesion.

Due to the contradictory results reported above, doubt must be raised as to the validity of such experiments using air-dried pretreated substratum since the character of the macromolecule on the surface is surely affected by the drying process.

Zaidi et al.[24] isolated high molecular weight materials associated with the bacterial extracellular matrix found on titanium and aluminum surfaces exposed to coastal seawater. This high molecular weight material was found to be immunologically related to surface adhesion-enhancing materials produced by surface-associated bacterial strains. Such specificity would imply that surface selectivity of a particular bacterial strain is based upon the interaction between these high molecular weight materials and the particular surface. Continued production of such immunologically specific material by the first colonizing bacteria will prejudice the biofilm surface for further specific colonization.

Imam et al.[25] further investigated the involvement of macromolecular cell surface components in cell interactions between the alga *Chlorella vulgaris* and several closely associated bacteria. The specific surface attachment between the *Chlorella* and glass and to specific bacteria is mediated by a lectin-like macromolecule associated with the cell surface. The specification of adhesion-enhancing materials shows that these molecules exercise a very selective effect determining which bacterial cells will adhere to *Chlorella vulgaris* in mixed culture. The role of these adhesion-enhancing materials in the subsequent specific colonization of a surface is apparent.

In a more practical study, LaMotta et al.[26] attempted to enhance the adhesion of slow-growing nitrifying bacteria to plastic surfaces and thus reduce the start-up times of biological fixed-film reactors by pretreating the surfaces with either a variety of commercially available polymers or bacterially derived exopolymers. Although naturally produced polymer enhanced initial biofilm development more than the synthetic polymers, their limited influence as an adhesion-promoter may not be economically feasible.

Very few workers have estimated the rates of molecular organic adsorption onto an inert surface because it occurs so rapidly and in such minute amounts. Table 2 summarizes the rate and extent of macromolecular adsorption estimated from literature data. The average time (t) required for a molecule to travel a distance (x) by molecular diffusion alone can be calculated from Equation 1,

$$t = x^2/2D \tag{1}$$

Assuming $x = 100\ \mu m$ and the molecular diffusivity $D = 0.5 \times 10^{-6}\ cm^2/sec$, the time required for an organic molecule to reach an inert surface is ~200 sec. At a bulk fluid organic concentration of 10 mg COD/ℓ, a molecular flux rate of 1.8 μg COD/cm^2/hr can be calculated. Data reported by Bryers[27] indicate an organic adsorption rate of 1.25 μg COD/cm^2/hr. Thus, it would appear that molecular diffusion alone could explain sufficiently the rapidity of macromolecular preconditioning of an inert surface.

2. Cellular Particle Transport

Cells, as particles suspended within a fluid, can be transported to a surface by any one, or a combination of, the following mechanisms:

- Motility (quiescent)
- Gravity (quiescent)
- Molecular diffusion (laminar flow)
- Eddy diffusion (turbulent flow)

Table 2
MAXIMUM RATE AND EXTENT OF MOLECULAR FOULING[27]

Maximum (nm/min)	Maximum accumulation (nm)	Maximum accumulation (μg COD/cm²)	Surface/fluid surface	Ref.
0.15—0.45	30—80	—	Pt[a]/w	12
0.004	7.1	—	Ge[b]/w	15
0.044	77.3	—	Ti[b]/w	15
0.01[c]	13.5[c]	1.2	Glass[d]	27
0.22[c]	22.5[c]	2.5	Glass[e]/lab	27

[a] Immersed in quiescent Chesapeake Bay water (3—4°C) containing 2.3 mg carbon/ℓ, salinity between 9—16%, and pH between 7.9—8.2.
[b] Gulf of Mexico water (22°C) flowing past the surface as fluid shear stress of 7.1 N/m^2. Salinity was 34%. Carbon concentration not reported.
[c] Estimated from measurements of chemical oxygen demand (COD) adsorbed per unit area. Assumed COD of protein is 0.855 mg COD/mg protein and protein density is 1.3 g protein/cm³.
[d] Medium consisted of a sterile 1:1 w/w of trypticase soy broth-glucose mixture (34°C; pH 9). The glass surfaces were immersed in tubes placed in a mechanical shaker. Carbon concentration was approximately 80 mg carbon/ℓ.
[e] Medium was effluent (30°C; pH 8) from a chemostat (10—20 mg/ℓ COD, 3 mg/ℓ polysaccharide) with no primary substrate remaining. Microorganisms were present (approximately 106 cells/mℓ) but no cells attached during the period of interest. Fluid shear stress was 3.8 N/m^2.

Based upon an estimate of the physical forces involved,[27] it is doubtful that motility, except over very small distances within quiescent systems, plays a significant role in cell transport to a surface.

Sedimentation is most likely the dominating mechanism for cell transport to a surface within a quiescent system. Assuming a spherical cell (diameter of 1.5 μm; a wet density of 1.001 g/cm³) suspended in a quiescent aqueous solution, the terminal settling velocity, estimated from Stokes law, is $\sim 1.2 \times 10^{-7}$ cm/sec. For a cellular suspension of 1×10^7 cells per cubic centimeter, the maximum possible flux of particles to a surface, due to sedimentation is 1.2×10^4 cells per square meter per second. Fletcher[28] reports, for a quiescent cell suspension of $\sim 1 \times 10^7$ cells per cubic centimeter, a net cellular accumulation rate of 0.5 cells per square meter per second. The discrepancy between estimation and observation suggests either that all the cells contacting the surface are not firmly bound and are lost during rinsing or that a significant portion of the suspended cells never reaches the surface (at the above settling velocity, it would take 12 hr for a cell to travel a distance of 50 μm).

Molecular diffusion may be the dominating mechanism for cell particle transport within a laminar flow situation. One can estimate the flux of particles to a surface exposed to laminar fluid flow from an equation proposed by Bowen et al.[29] Particle flux rates for shear stresses of 0.8 and 0.09 N/m^2 are calculated at 9.2×10^5 and 4.5×10^{-5} cells per square meter per second, respectively. Powell and Slater[30] report net cellular accumulation rates in laminar flow reactors at a cell concentration of 1.4×10^7 cells per cubic centimeter of 3.3×10^{-4} and 2.2×10^{-3} cell per square meter per second for the same shear stresses, 0.8 and 0.09 N/m^2, respectively. Theory implies in laminar flow that an increasing shear stress will increase particle flux. Powell and Slater report a decreasing net cell accumulation rate with increasing shear stress which could be attributed to either a shift from molecular diffusion as the dominating mechanism or a decreasing cellular attachment rate with increasing shear stress.

Beal[31] describes particle flux in a turbulent flow system using the equation

$$N = (D + D_e)\ dC/dy \tag{2}$$

where N = particle flux (cells/L^2/t), D = molecular diffusion coefficient (L^2/t), D_e = eddy diffusion coefficient (L^2/t), C = particle concentration (cells/cm^3), and y = distance to surface (L). D_e is dependent upon the turbulent intensity in a fluid and thus varies with the distance from the surface. D_e values are generally orders of magnitude greater than D. Beal integrated Equation 2 to give

$$N_{wall} \cong K_D\ C_{AVG}$$

$$\text{with} \quad K_D = K\ V_nP/(K + V_nP) \tag{3}$$

where K_D = deposition coefficient (L/t), V_n = velocity component normal to the surface (L/t), K = transfer coefficient (L/t), and P = "sticking" efficiency (dimensionless). The K in Beal's model is a complex function of particle physical properties and prevailing fluid flow parameters for fluid flow in a circular tube. Beal's work indicates that particle transport rate is directly proportional to bulk particle concentration. Bryers and Characklis[32] describe experiments measuring cellular deposition in a circular tube which indicate that for a fivefold increase in cell mass concentration (4.0 → 20 mg DW/ℓ), deposition rate increased by a factor of 4.5 (fluid shear stress constant). The effect of changing fluid flow is not as simple, since K is a complex function of fluid and particle conditions. Increasing fluid velocity could increase or decrease K and decrease the mass transfer boundary layer, thereby increasing transport rate to the surface.

Beal's equations predict the particle transport coefficient, K_m, to increase with fluid velocity within the ranges considered by Bryers and Characklis.[32] However, their results indicate a net cellular accumulation rate due to deposition only which was unaffected by doubling fluid velocity. Once again, this discrepancy between particle transport estimates and observed deposition rates arises since "deposition" includes not only cellular transport but adhesion and detachment rates. Therefore, cell transport in turbulent flow may increase with increasing flow rate but, for example, cell adhesion rate may decrease with increasing shear. Duddridge et al.,[33] in adhesion studies with *Pseudomonas fluorescens*, also concluded that increasing shear stresses decreased the efficiency of particle adhesion.

3. Adhesion

Once at a surface, bacteria can adhere either reversibly or irreversibly. The principal criteria for reversible adhesion is that cells can be easily removed by either washing[34] or, if motile, can swim along and away from the surface. Should reversible adhesion hold a cell at a surface for sufficient time, then the cell-surface binding can "mature", negating simple removal procedures. Irreversible adhesion is often associated with the production of extracellular polymers (EP).[4,34-36] Fletcher[37] employs a somewhat different nomenclature referring to passive attachment which arises due to weak short-range physicochemical adhesion, while active attachment results due to physiological activities of the cell. It is not clear whether EP is produced before or after the cell reversibly adheres to the surface. Fletcher and Marshall[10] and Sutherland[38] review the literature concerning the main environmental factors (e.g., nutrient type and concentration, cellular growth rate, carbon vs. nitrogen limitation, temperature, cation concentration, pH, cell species) affecting polymer production by surface-association cells. Other mechanisms of cell-surface adhesion include cellular appendages, e.g., holdfasts, stalks, and pili.[39]

Most adhesion studies are carried out in quiescent systems; thus, adhesion rates estimated from such results may be particle transport-rate limited and may not be applicable to laminar or turbulent flow systems. Using a capillary glass flow cell, Powell and Slater[30] observed the deposition and eventual adhesion of *Bacillus cereus* in laminar flow (0.4 > Reynold's number > 16) onto a clean glass surface. They noted that reversible adhesion to the wall was often incomplete, with cells contacting the wall, then rolling or sliding along the surface before either attaching firmly or being re-entrained into the fluid. An analysis of the contact time which cells spent at the surface indicated a logarithmic distribution skewed toward short contact times (<10 sec).

No such evidence of reversible adhesion has been reported for turbulent flow conditions.

Mathematical descriptions of rates of adhesion, based upon the Derjaguin-Landau-Verwey-Overbeek (DLVO) theory or extensions of the DLVO theory[40] can be successfully applied to reversible adhesion to interpret observed responses in cellular adhesion to changes in ionic strength, Ca^{2+}/Mg^{2+} concentrations and substratum interfacial physics. However, application of such particle dynamics theories to situations of active or irreversible adhesion is open to question due to complications created by cellular metabolic activity (i.e., EP production; biofilm surface chemistry differences). Mathematically, physiological effects on cell adhesion would require complex models. For example, Ca^{2+} and Mg^{2+} concentrations are known to greatly enhance cellular adhesion rates.[41] However, such enhancement could be due to the divalent cation effects on bridging or crosslinking of polymer side-chains, the electric double layer, or the physiological activity of the cells.

Consequently, most researchers opt for measuring the net accumulation of cells at a surface, i.e., deposition rate. Powell and Slater[30] derive a cell balance on the surface of glass rectangular conduits of the form

$$dN/dt = J_d + [\mu_s - (\Psi_o)] \quad (4)$$

where N = number of cells irreversibly attached to the surface (cells/cm^2); J_d = particle flux rate to the surface (cells/cm^2/hr); μ_s = cellular growth rate (hr^{-1}); Ψ_o = removal rate of irreversibly attached cells (hr^{-1}), which is a function of shear stress at the surface, τ_o (N/m^2). Powell and Slater estimated J_d values from experimental observations of *B. cereus* depositing on glass under laminar flow conditions using Equation 4. Experimentally estimated values of J_d were considerably less than predicted by the particle deposition theory of Bowen et al.[29] for laminar flow conditions.

Bryers and Characklis[32] report net deposition rates of a mixed bacterial culture (average cell diameter about 3 μm) determined within circular tubes under turbulent flow conditions. Table 3 indicates that the cellular deposition rate was directly proportional to the concentration of suspended particulates, as predicted by deposition theory according to Beal[31] for turbulent flow. However, when the Reynolds number was doubled, an increase in deposition rate was not observed, contrary to particle deposition theory.

Data from Powell and Slater[30] and Bryers and Characklis[32] imply a cellular "sticking efficiency" of less than 1.0, since observed deposition rates are less than predicted by particle transport theories (which assume 100% of particle-surface contacts result in adhesion). Characklis[42] estimates that cellular sticking efficiency is proportional to shear stress as indicated in Figure 7.

Baltzis and Fredrickson[43] model the accumulation of cells on a surface as the balance between attachment and removal processes,

$$dN/dt = r_a - r_r \quad (5)$$

where the rate of attachment is assumed to be a Langmuir function of the amount of surface available for cell attachment, i.e.,

Table 3
RATES OF BACTERIAL DEPOSITION TO GLASS TUBES UNDER TURBULENT FLOW CONDITIONS[32]

Reynolds no.	Suspended biomass (mg/DW/ℓ)	Deposition rate (μg COD/cm²/hr)
13,000	20.0	1.1—1.2
26,000	20.0	0.8—1.0
13,000	4.0	0.30

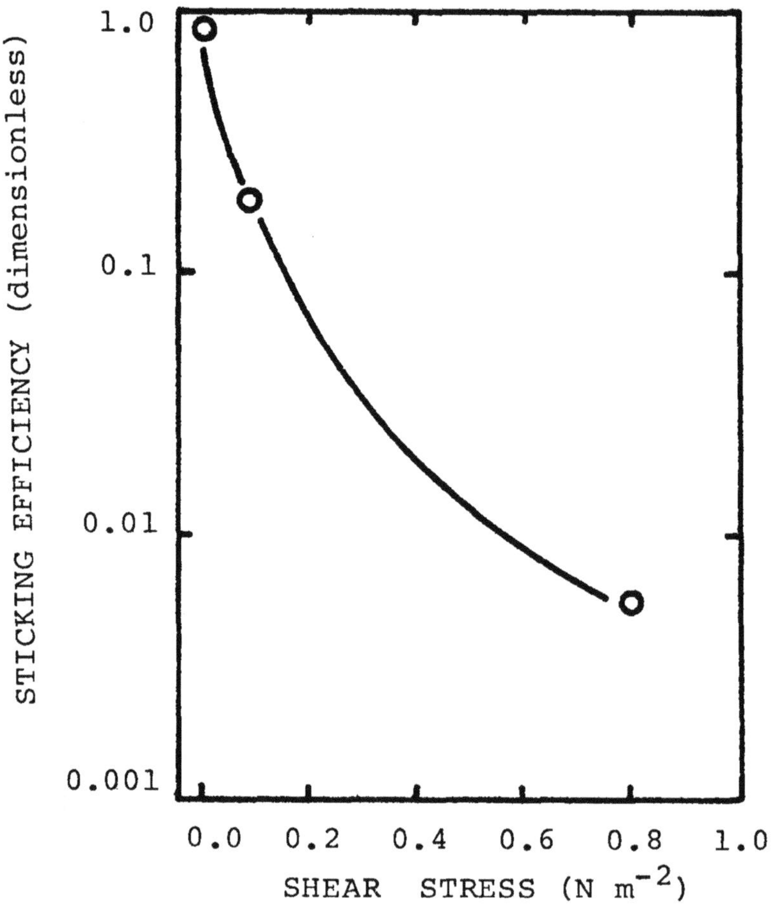

FIGURE 7. Influence of fluid shear stress on adhesion efficiency of bacterial cells.[42]

$$r_a = kX_{bulk}(N_{max} - N) \tag{6}$$

where X_{bulk} = bulk fluid concentration of suspended cells (cells/ℓ); N_{max} = maximum surface area occupied by cells due to adhesion only (cells/cm²); N = number of cells attached at some time t (cells/cm²), and k = the adhesion rate constant (ℓ/cells/sec). The authors provide no experimental corroboration for this particular rate expression. N_{max} is probably a characteristic parameter of the surface and microbial species; actual values of N_{max} may be much less than the total surface monolayer coverage, as once expected. A study by Dabros and van de Ven[44] found that negatively charged latex particles adhered to negatively charged glass, oscillated about a given point, and excluded from subsequent adhesion an area 20 to 30 times their geometric diameters. Consequently, N_{max} may correspond to only 1 to 3% of the total surface coverage which corresponds well to literature values for cell adhesion in nongrowing systems.[45]

B. Biofilm Metabolic Processes

Cellular growth, replication, substrate conversion, endogenous decay, and extracellular polymer production are collectively those metabolic processes euphemistically termed "biofilm growth". Unfortunately, the term "biofilm growth" is all too often confused with "net biofilm accumulation"; the latter phenomenon being composed of several contributory mechanisms of which "growth" is but one.

1. Substrate Conversion for Growth and Replication

The growth and replication of cells trapped within a biofilm can be described using an autocatalytic function of cell concentration similar to those used for suspended cell situations, i.e.,

$$R_B^G(t) = \mu B(t) \tag{7}$$

where R_B^G = local area growth rate of attached cells ($M_x/L^2/t$); μ = local specific growth rate constant (t^{-1}), and B = area concentration of attached cells (M_x/L^2).

Biofilm "growth" kinetics differ from those of suspended cell systems in that surface-bound cells, producing biofilm exopolymer, may possess a different cellular stoichiometry than suspended cells,[46,47] and the specific growth rate, μ, is a function of the local substrate concentration which is dictated not only by the biological reaction rate, but also by mass transfer resistances. In the case of multiple species biofilms (discussed in Section IV), even the concentration of bound cells may be a function of position as well as time, i.e., B(x,t).

Description of biofilm substrate removal kinetics has received perhaps the most mathematical attention of all the biofilm processes discussed here. Design of wastewater fixed-film treatment reactors has created the need for models describing the overall substrate removal rate (per reactor volume) by a certain amount of biofilm. Substrate removal rate, r_s, is simply a stoichiometric ratio of the cellular biofilm growth rate,

$$r_s = R_B^G \, A/Y_{B/S} \, V = \eta \, \mu \, BA/Y_{B/S} \, V \tag{8}$$

where A = reaction surface area (L^2); V = reaction volume (L^3); $Y_{B/S}$ = stoichiometric ratio of biomass produced per substrate consumed (M_x/M_s); η = biofilm effectiveness factor. η is a dimensionless parameter which compensates the substrate removal rate for possible mass transfer effects. Defined as the ratio of substrate removal with mass transfer to the maximum substrate removal without mass transfer, η has numerical values between $0 \rightarrow 1$ and is a complex function of biofilm geometry, reaction kinetics, substrate concentration, and substrate diffusivity. Perhaps the most overworked subtopic within biofilm formation literature is the derivation of a functional expression for η under specific reaction situations.[48-57]

Unfortunately, most biofilm substrate kinetic literature is based on the assumption that biofilm thickness (or mass) is constant (i.e, dB/dt = 0) although substrate is continually consumed. Researchers assume (for mathematical simplicity) that biofilm production due to growth is either negligible or counterbalanced by processes reducing or detracting biofilm, e.g., biofilm removal or endogenous decay. Such models, instructive but now redundant, have shed little light on biofilm formation kinetics and are not applicable to transient growth conditions (although some imply the opposite!).[58,59]

Expressions for η in terms of known reaction parameters are traditionally derived from a steady-state substrate balance over an elemental volume of a known biofilm thickness. The resulting differential equation is

$$D(d^2S/dx^2) = -r_s\ A/V \tag{9}$$

where S = substrate concentration (M/L^3) and x is the direction of substrate transport. Substrate profiles within a biofilm are determined by integration of Equation 9 once appropriate boundary conditions and the dependency of r_s on substrate concentration are specified. Once the substrate profile is known as a function of x, the overall substrate flux to the biofilm, N_s ($M_s/L^2/t$), can be estimated from Equation 10.

$$N_s = -D\ dS/dx|_{x=L} \tag{10}$$

where L = total biofilm thickness (L). An expression for η is merely the result of Equation 10 divided by the maximum rate of substrate removal if the entire thickness L of biofilm were exposed to the bulk fluid concentration of substrate, i.e.,

$$\eta = \frac{-D(dS/dx)|_{x=L}}{-(r_sA/V)_{S=S_{bulk}}} \tag{11}$$

Grady[60] provides the most comprehensive review concerning biofilm substrate kinetics. A summary of the more common rate expressions employed in biofilm kinetics is given in Table 4.

When used in material balances, most of these agree well with observation; however, this agreement is, in a sense, fortuitous considering the number of oversimplifications made in most derivations. These include the following;

1. Biofilm thickness, L, is constant during substrate consumption.
2. Biofilm density, ρ_B (M_x/L^3), is also assumed constant and uniform for the entire film.
3. Concentration of biofilm cells, B (x,t), in Equation 7 is typically replaced with the observed biofilm density, ρ_B, which overestimates the active cell mass in a biofilm.
4. Mass transfer of limiting substrate is by molecular diffusion only, which may not be true for biofilms with highly fibrous surface morphologies.[80]

2. *Endogenous Decay, Death, and Lysis*

Concepts of cell death, lysis, and maintenance requirements are either ignored in most mathematical treatments of biofilm "growth" or considered collectively as one lumped process called "decay". Decay in suspended cultures can imply the loss of cell mass (or a change in the oxidation state of cellular material) due to either a prolonged absence of an exogenous substrate or a constant energy requirement by the cell to maintain certain metabolic activities. In steady-state biofilm growth kinetics, however, decay is all processes that counterbalance production of biofilm mass, thus keeping biofilm mass constant.[55,56,65] Unfortunately, this approach erroneously lumps such processes as biofilm shear-removal, extracellular polymer production, lysis, sloughing, and predation under the heading of decay, thus overestimating the decay constant throughout the development of a biofilm. Should biofilm thickness (or mass) remain constant in an engineered system it would most likely result from hydrodynamic forces, not decay processes. Unrealistically high values for biofilm cell decay constants would be expected if the lumped concept of decay were employed. If

Table 4
CHARACTERISTICS OF BIOFILM RATE EQUATIONS[60]

Model no.	Type of reaction rate expression[a]	Limiting material[b]	Consideration of nonlimiting material utilization?[c]	Cell decay in electron acceptor balance?[c]	Film thickness[d]	Distinction between thick and thin films?[e]	Ref.
1	SSM	ED	No	N/A	Input	Yes	49,61
2	SSM	ED	No	N/A	Input	N/N	62,63
3	SSM	ED	No	N/A	Output	N/N	64
4	SSM	ED	No	N/A	Output	Yes	54
5	SSM	ED	No	N/A	Input	Yes	59
6	SSM	ED	No	N/A	Output	Yes	65
7	SSM	ED	No	N/A	Input	Yes	66
8	SSM; SSB	ED	No	N/A	Input	N/N; No	67
9	SSB(1)	ED	No	N/A	Output	N/N	48
10	SSB(1)	ED	Yes	No	Output	N/N	68
11	SSB	EA	No	No	Input	Yes	69
12	SSB(O)	EA	No	No	Input	Yes	70,71
13	NDSM	ED, EA	N/A	No	N/N	Yes	51
14	NDSM	ED, EA	N/A	No	Input	Yes	72,73
15	NDSB(0)	ED, EA	N/A	No	Input	Yes	74
16	IDSM	ED, EA	N/A	No	Output	Yes	75
17	IDSM	ED, EA	N/A	No	Input	N/N	76
18	IDSM	ED, EA	N/A	Yes	Input	N/N	77—79

[a] SSM, single-substrate, Monod; SSB, single-substrate, Blackman, (0) = zero order only, (1) = first order only; NDSM, noninteractive double-substrate, Monod; NDSB(0), noninteractive double-substrate, Blackman, zero order only; IDSM, interactive double-substrate, Monod.

[b] ED = electron donor; EA = electron acceptor.

[c] N/A means that the question is not applicable to the particular reaction rate expression employed.

[d] Is the value of the film thickness an input or an output? N/N means that it is not necessary to know the thickness to proceed.

[e] N/N means that it is not necessary to make a distinction to solve the equations employed.

maintenance and shear removal are treated individually, quite realistic values for decay constants are estimated (0.045 to 0.14/day at 37°C[27,32]).

Endogenous decay is traditionally incorporated into a net biomass turnover rate, r_{net} ($M_x/L^2/t$), expression, i.e.,

$$r_{net} = r_g - k_e B = (\mu - k_e) B \tag{12}$$

where k_e = first order biomass decay constant (t^{-1}). Note in most biofilm models B (the concentration of active cell mass) is equated to the total biofilm density, ρ_B. This substitution tacitly implies that all the biofilm mass participates in substrate removal, i.e., all the cells are 100% active and the extracellular polymer component is metabolically active.

3. Extracellular Polymer Production

Biofilms, as evident from electron micrographs (Figures 2 to 5), are not made up of microorganisms stacked in successive layers, but rather consist of cells entrapped within a gelatinous matrix of extracellular polymer (EP). EP is produced directly from the surface-associated microorganisms. Most researchers believe EP acts as a "cementing" agent to reinforce cell binding to a surface; although recently, physical chemists have suggested that EP production may be a cell's attempt to "free" itself from a surface.[81] Further research now indicates that EP is not a passive gel matrix but rather exhibits physical, chemical and electrical responses to environmental stimuli (e.g., changes in pH, substrate type and concentration, and temperature) and these responses may be under indirect control of the bound cells.[82]

Only recently has EP production been taken into account when estimating the stoichiometry and kinetics for biofilm-mediated process. Kissel et al.,[83] in a structured model of mixed culture bacterial biofilm formation, simulate the continuous production of an "inert" portion of biofilm. The authors assume that the production rate of inert material is directly proportional to the decay rate of viable cell mass. It is not clear whether the authors are considering EP material plus inert cell mass in this inert fraction or, more likely, only cellular material that is no longer metabolically active (i.e., dead cells). Since decay processes will prevail within the lower depths of a thick biofilm, the Kissel model inherently produces more inert material at lower depths which contradicts observations.

Trulear[46] presents another more structured model of a pure culture bacterial biofilm which incorporates specifically EP production. Contrary to the Kissel model, Trulear simulated the rate of EP production using a Leudeking-Piret rate expression,

$$(r_{EP})^T = (r_{EP})^g + (r_{EP})^{NG} \tag{13}$$

where $(r_{EP})^T$ = total rate of EP production ($M_{EP}/L^2/t$), $(r_{EP})^g$ = growth-associated EP production rate ($M_{EP}/L^2/t$), $(r_{EP})^{NG}$ = nongrowth associated EP production rate ($M_{EP}/L^2/t$). Details concerning this structured model are provided in Section III.B.1.

C. Biofilm Removal Processes

Biofilm can be removed from a surface in any one of the following ways: (1) predator harvesting, (2) abrasion, (3) shear related detachment, (4) sloughing, and (5) human intervention. Current knowledge and mathematical description of these processes, except for (5) will be discussed in the subsections to follow.

1. Predator Harvesting

Grazing or harvesting of a biofilm by protozoa is a common occurrence, but one biofilm removal process which has been completely ignored within mathematical models. One simple

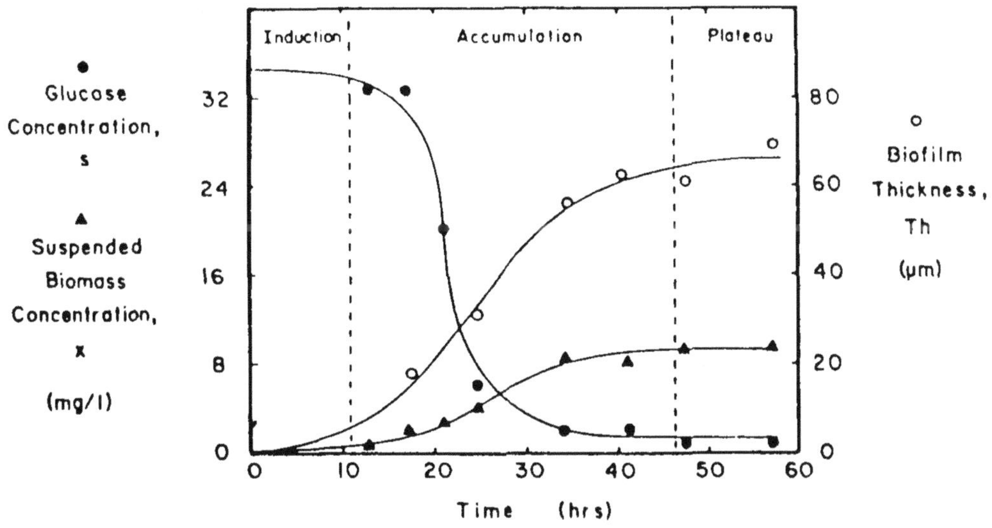

FIGURE 8. Effluent glucose and suspended biomass concentration and biofilm thickness in a mixed culture grown continuously within a rotating annular reactor.[84]

rate expression for biofilm removal by predation would be an overall second-order function, first order in both predator and biofilm concentrations. Unfortunately, experimental studies involving bacterial biofilms-protozoan communities do not exist.

2. Shear-Related Removal

Several workers have shown that biofilm material is removed continually throughout biofilm development.[27,84] Figure 8 indicates that for chemostats operated at dilution rates in excess of the culture's maximum growth rate, suspended biomass is observed to parallel biofilm development. Results of each study show significant suspended biomass is sheared continuously from the biofilm and biofilm removal rates are proportional to biofilm amount up to a certain thickness. Rittmann,[85] in a rather prodigous stretch of imagination, describes biofilm removal rate data of Trulear[84] using an expression that is first order in biofilm amount (solid line, Figure 9). However, experiments indicate that removal rate is not such a simple function, but rather highly dependent upon, among other things, the hydrodynamics of the system. Once biofilm thickness exceeds the laminar sublayer in a system, shear stress at the same velocity increases dramatically which, in turn, increases the biofilm removal rate.[32] Consequently, biofilm removal rate expressions should specify certain experimental bounds.

Zelver[86] reports that biofilm shear removal, in response to a sudden increase in shear, is also a function of the fluid shear prevailing during steady-state growth. Biofilm grown at low shear stress is more readily removed by an increase in shear stress than is a biofilm grown at a high shear stress.

3. Abrasion

Biomass captured within stainless steel spheres or polyester sponge matrices can be employed in fluidized bed biological reactors for either biotechnological purposes or waste-water treatment.[87] Biomass develops within the interstices of such support particulates which are hydraulically fluidized by influent liquid. Mixing and agitation cause the particles to collide and the resulting abrasion continuously removes excess biomass. Consequently, a constant biomass hold-up, in particles of known size and shape, is maintained. Thus, at a steady-state biomass holdup, the removal rate would equal the net biomass production rate; however, there seem to be no mathematical models of biofilm formation that incorporate abrasion mediated removal.

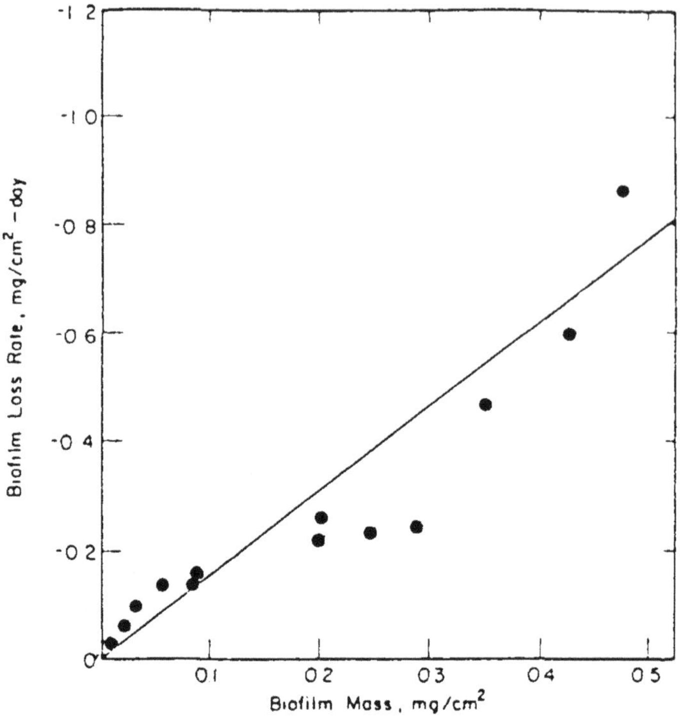

FIGURE 9. A linear correlation between biofilm loss rate and biofilm mass drawn by Rittmann[85] based upon data from Reference 84.

4. *Sloughing*

Sloughing is an event where large amounts or entire sections of biofilm leave a support surface and enter the surrounding media. Unlike shear-related removal, sloughing is a periodic occurrence. There are no clear explanations for sloughing, most likely because such events have a multitude of system-specific causes.

Sloughing has been linked to the occurrence of anaerobic activity within the depths of very thick, but otherwise aerobic biofilms. Due to oxygen transfer limitations in such cases, deep layers of biofilm become anaerobic, producing volatile acids, decreasing pH, and perhaps producing insoluble gases which may weaken the biofilm structure; however, this premise does not explain the sloughing of strictly anaerobic biofilms as recorded in Figure 10.[88]

Annual cycling of trickling filter biofilm thickness reported by Hawkes and Shephard[89] and Heulekian and Crosby[90] were also confirmed in laboratory experiments where the total weight of a filter was measured over a 2-year period. Data indicate a major seasonal slough (amplitude ~50% of total winter mass) each spring and a minor monthly sloughing cycle (amplitude 10 to 25% of winter weight).

The first documented attempt to mathematically incorporate sloughing in a biofilm formation model was made by Howell and Atkinson.[64] Their approach was purely an empirical model of sloughing events and provided no biochemical or microbiological mechanism of sloughing. Biofilm grew at a rate proportional to the overall substrate flux which could be either reaction-limited, mass transfer-limited or both, depending upon prevailing reaction conditions. At some point in the development of the biofilm, thickness was sufficient to deplete the limiting substrate to zero at the biofilm-inert surface interface. At this time, biofilm thickness was computationally "reset" to a very small value, the "computer-sloughed" biomass uniformly distributed in the liquid and biofilm allowed to develop anew. Howell and Atkinson thus linked sloughing to limiting substrate depletion within the biofilm.

FIGURE 10. A pure culture of anaerobic bacteria degrading NTA in continuous culture. Note biofilm, after 2 months, experienced a localized sloughing randomly over the vessel internal surfaces.[88]

Harremoës et al.[91] present a 4-year study of a denitrifying fixed-film pilot plant which experienced operational problems due to sloughing associated with nitrogen bubble formation. Bubbles created localized biofilm sloughing as seen through the transparent walls of the laboratory reactor (Figure 11). In addition, a rapid headloss in the trickling filter was attributed to bubble formation. Once reactor liquid becomes N_2 saturated, nitrogen bubbles will form in the biofilm which places a practical limit to the NO_3-N removal rate possible in a fixed film reactor. Similar sloughing is expected to occur in anaerobic systems producing methane and CO_2. Bubble formation and associated sloughing could be minimized in a fluidized bed fixed-film reactor since film thicknesses are less and the agitation may physically dislodge fine bubbles before they coalesce and disrupt the biofilm structure.

Arvin and Kristensen[92] show significant pH gradients can exist in very minute denitrifying biofilms. Such gradients could lead to adverse pH conditions in the depth of a biofilm even though bulk liquid conditions are maintained physiologically tolerable. The authors also report the precipitation of significant amounts of calcium phosphate in the biofilm as a result of the observed pH gradients which could also contribute to sloughing.

Bakke and Characklis[93] monitored substrate uptake, soluble product formation, biofilm

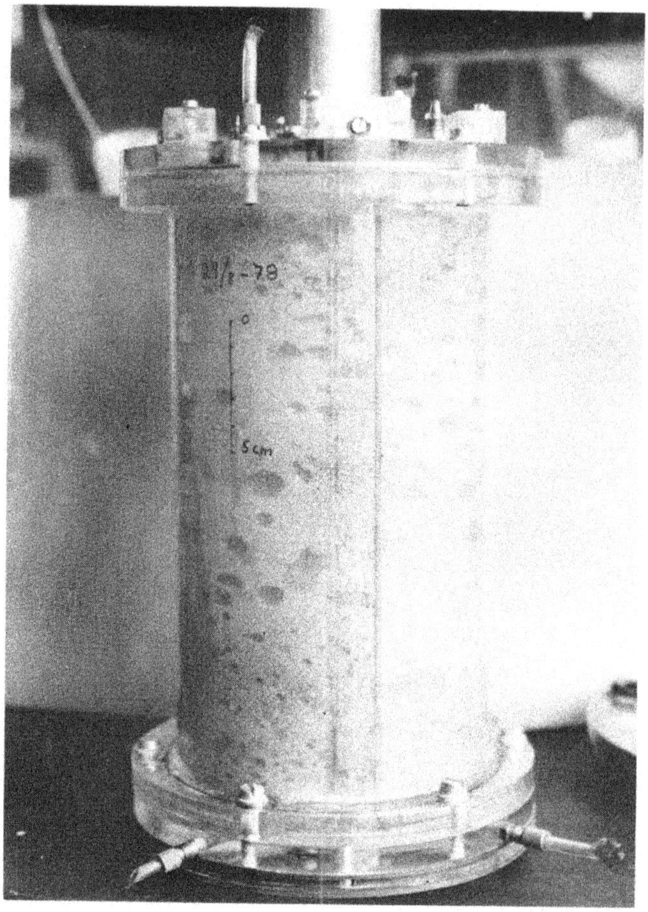

FIGURE 11. Photograph illustrating biofilm sloughing due to nitrogen bubble formation within the biofilm created by supersaturation of nitrogen in the bulk liquid.[91]

EP concentration, bulk liquid cell concentration, and bulk liquid EP concentration before and after a step increase in influent substrate concentration. Substrate flux increased immediately upon the increase in substrate load. Until the biofilm adjusted to the richer environment, substrate concentrations increased in the bulk liquid. An immediate increase in effluent soluble product and "lagged" increases in biofilm cellular reproduction and EP production were also observed. More surprisingly, significant EP detachment, not paralleled by cellular detachment, occurred within the first minutes after a transition. Bakke and Characklis suggest the EP detachment indicates that EP may serve as a biofilm diffusion regulator under passive control of the bound bacteria. According to their model, bacteria, in responding to increased substrate transport, moderate their membrane electrical field perhaps creating local pH conditions that release bound EP into the surrounding liquid. To the observer, such an event would appear to be sloughing.

Unfortunately, no in-depth study of biofilm sloughing exists even though information on such natural events may lead to more efficient artificial control measures.

III. MATHEMATICAL DESCRIPTION OF BIOFILM FORMATION

Models which describe microbial growth either as an increase in total biomass or which consider only simple, single microbial conversion reactions are called *unstructured models*. Such lumped parameter approaches do not detail, for example, mixed culture population dynamics or changes in cellular composition in response to changes in environmental conditions. Actually, an unstructured model, based upon total cellular mass, cannot distinguish between pure and mixed cultures. These models, although relatively simple, have been applied throughout the industrial fermentation industry, the biological treatment sector, and in microbial ecology for purposes of data interpretation, design, control, and optimization. Since they are based upon a pseudo-steady-state approximation, unstructured models are best suited for describing microbial systems under uniform conditions which foster "balanced growth".

Conversely, *structured models* do consider as a function of prevailing conditions the additional detail of mixed culture population dynamics, microorganism composition, and/or multiple reaction schemes. Consequently, structured models are inherently suited for simulating microbial physiology under transient, nonsteady conditions.

In the following section, both modeling philosophies will be applied to biofilm formation; Section III.A exemplifies the more common unstructured biofilm formation models, while Section III.B illustrates in some detail two different structured biofilm models. For more details concerning the theory and practice of both structured and unstructured models, the reader is directed elsewhere.[94]

A. Unstructured Models of Biofilm Formation

1. Basic Assumptions

The model derived below will summarize the more common points in most existing biofilm accumulation models. In an unstructured approach, based for example upon total mass, such models can be applied equally to either pure or mixed cultures, provided, in the latter case, that culture-averaged kinetic and stoichiometric parameters are used.

Aside from a few exceptions, all biofilm models reviewed in this section simulate biofilm accumulation assuming ideal completely mixed reactor behavior, thus tacitly neglecting spatial heterogeneity in the bulk fluid. Referring to the system and nomenclature in Figure 12, assumptions used in the mathematical development are

1. Biological activity occurs within the reaction volume, V, of a completely mixed reactor. Biofilm formation occurs uniformly over a surface area, A.
2. Suspended biomass concentration, C, in the bulk fluid arises due to either suspended biomass growth or to shear-related removal of surface-bound biomass. Suspended biomass is removed from the reactor by effluent leaving the reactor and by cell-particle deposition onto reactor surfaces. Note that a material balance on suspended biomass can be based upon any reasonable measure of biomass concentration (e.g., cell number, cell mass, total suspended solids, suspended cellular carbon, etc.) provided the relationship between changes in that measured parameter and the limiting substrate used is known.
3. A sterile, growth-limiting substrate, S, enters the reactor at an influent concentration S_i. Substrate is consumed by both suspended biomass (a homogeneous reaction) at a rate defined as R_S^C and by surface-bound biomass (a heterogeneous reaction).

2. A General Unstructured Model

Within the confines of these three limitations, material balances for suspended biomass and limiting substrate can be written simply as

Suspended biomass balance

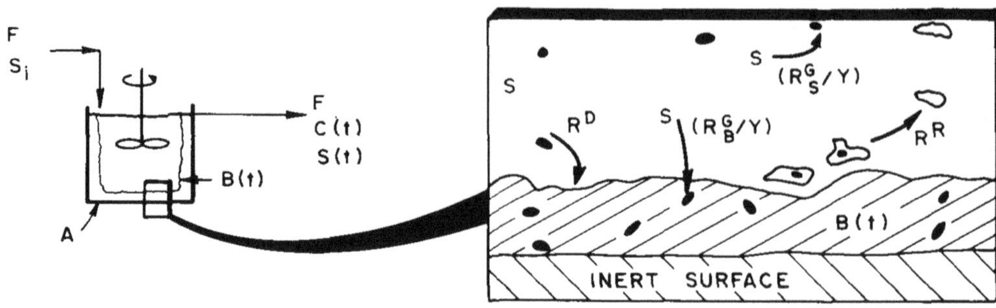

FIGURE 12. Hypothetical process analysis employed to model pure and mixed culture biofilm effects on chemostat dynamics.[95]

$$VdC/dt = -FC + R_S^G V + R^R A - R^D A \tag{14}$$

Limiting substrate balance

$$V\ dS/dt = F(S_i - S) - [R_S^G V/Y_{C/S}] - [R_B^G A/Y_{B/S}] \tag{15}$$

Definition of the various process rates (i.e., R_S^G, R^R, R^D, R_B) are given in Section II and reiterated in Figure 12. Several of these rate processes depend directly on changing biofilm concentration; consequently, an auxiliary equation describing the change in biofilm concentration with time is required.

Biofilm accumulation at a surface is assumed to proceed according to the scenario presented in Section II. Both biofilm growth and cell deposition processes positively contribute to, while shear-related biofilm removal acts to retard biofilm formation (the reader should be aware that older articles concerning biofilm formation may have considered growth as the only pertinent process and thus kinetic parameters and conclusions resulting from these works may need reexamination). Thus, biofilm accumulation can be described by the equation

Biofilm accumulation equation

$$A\ dB/dt = AR^D + R_B^G A - R^R A \tag{16}$$

Equations 14 through 16 form a very simple unstructured model of biofilm formation within a completely mixed reactor. Since Equations 14 to 16 are nonlinear, ordinary differential equations coupled through the various rate expressions, they require either simultaneous, numerical integration (for the unsteady-state situation) or simultaneous algebraic solution (under steady-state conditions, i.e., $dS/dt = dC/dt = dB/dt = 0$). Section II discusses the various functional expressions reported for the different rate processes in Equations 14 to 16.

Bryers,[95] employing the functional expressions, operating conditions, and numerical constants given in Figure 13, solves these three equations for the case of pure culture (*Pseudomonas putida*) biofilm formation in a fermenter. Figure 13A and B illustrates the relatively good agreement between simulation and experimental observations.

Trulear,[84] with a similar set of equations, determined the various pertinent rate expressions involved in mixed bacterial culture biofilm formation within a rotating annular reactor. Typical results for biofilm growth and shear removal are shown in Figure 8.

For conditions of a mixed culture well-mixed biofilm reactor operated at D = 4.0/hr to eliminate suspended biomass growth, Figure 14 illustrates a perturbation technique which Bryers and Characklis[32] employed to determine the individual rate constants for the various processes included in Equation 16.

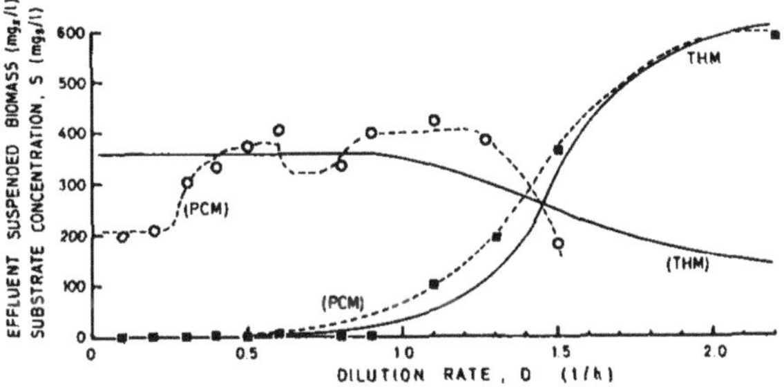

Steady-state effluent substrate (■) and suspended cell mass (○) for *Pseudomonas putida* grown in a chemostat at $S^o = 1.0$ g_s/L asparagine as reported in ref. 95. The symbol (——) indicates predictions of Topiwala-Hamer model and (-----) indicates predictions of a dynamic model (ref.95). Growth conditions in Table below.

A

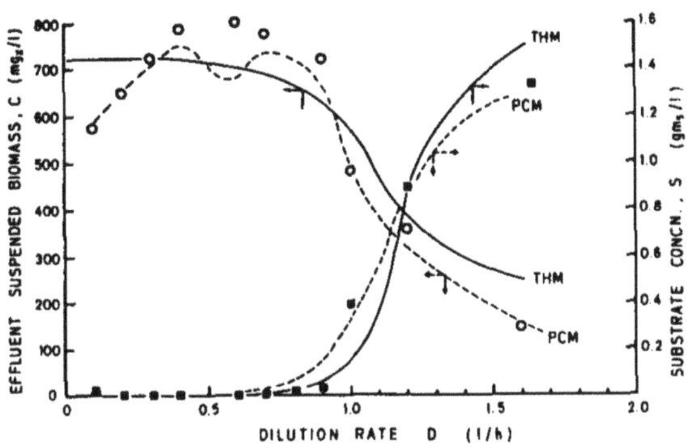

Same steady-state effluent data as above for *Pseudomonas putida* grown in a chemostat a $S^o = 2.0$ g_s/L. Symbols as above. Growth conditions in Table below.

B

Parameters used in either the dynamic pure culture model or the steady-state Topiwala-Hamer model to simulate *Pseudomonas putida* reults (ref.95). Dilution rates varied from 0.1 to 2.2 h^{-1}.

Parameter	A (cm^2)	V (cm^3)	$\hat{\mu}$ (h^{-1})	k^R (h^{-1})	k^R ($L/mg_x h$)	K_s (mg_s/L)	M_N (mg_x/cm^2)	M^* (mg_x/cm^2)	Y and α (mg_x/mg_s)
Value	800	970	0.59	0.2	3.0×10^{-5}	10.0	0.50	5×10^{-4}	0.32

FIGURE 13. Predictions of biofilm model[95] in comparison to experimental results. Model parameters summarized in the table above.

B. Structured Models of Biofilm Formation

Those few structured models of biofilm formation that exist have only considered either cellular and extracellular component changes during formation of a pure culture biofilm or population dynamics during mixed culture biofilm formation. Each of these structured approaches will be discussed in some detail in the sections to follow.

1. Substrate Metabolism in a Pure Culture Biofilm

Consider the substrate "destinations" in the unstructured biofilm formation described in Section III.A. Substrate is transported from the bulk liquid phase through the biofilm and consumed for what was termed "biofilm growth". For lack of information or insufficient incentive, most biofilm models employ an unstructured view and ignore the various metabolic processes comprising biofilm "growth". However, to establish the actual metabolic activity of biofilm-bound cells, a more detailed accounting of the various uses of substrate is needed.

Trulear[46] and Bakke et al.[47] developed the only current mathematical model that accurately portrays substrate metabolism in a pure culture biofilm developing within an annular rotating biofilm reactor (Figure 15). Here, a pure culture (*Pseudomonas aeruginosa*) biofilm consumes glucose as the sole carbon and energy substrate for the following metabolic processes: cellular synthesis and reproduction (collectively termed "cellular growth"), extracellular polymer production, and maintenance. All material balances are based upon total organic carbon and assume ideal completely mixed reactor behavior. Building on the previous unstructured scenarios of biofilm formation, the model of Trulear and co-workers assumes biofilm total carbon is the net result of the metabolic processes above and a shear-related removal process with deposition ignored. Within these bounds, their model comprises general unsteady-state balances for substrate carbon, extracellular polymer carbon, and cellular carbon in both the bulk fluid and the biofilm as summarized in Table 5*. Cellular growth processes consume substrate at a rate that is described by a Monod-like expression of substrate concentration. Extracellular polymer production rate is described using the growth and nongrowth associated rate approach of Luedeking and Piret.[96] Detachment processes are modeled using a simple power law rate expression of biofilm quantity. Comparison between experiment and model is given in Figures 16A and B.

2. *Population Dynamics during Mixed Culture Biofilm Formation*

Although most biological wastewater treatment processes and natural environments spawn biofilms of mixed bacterial populations, most mathematical treatments, due to their unstructured nature, have modeled biofilm development as a pseudo-pure culture system, measuring accumulation as an increase in biofilm mass, thickness, or total biofilm organic carbon. In this same context, most biofilm substrate kinetic models have assumed only one substrate is limiting and that only one microbial conversion process (e.g., carbon oxidation, nitrification, denitrification, methane production) occurs. Such assumptions clearly do not relate to the real systems cited above since each conversion is itself a sequence of biological reactions involving mixed populations, and frequently, these multiple conversion processes occur simultaneously in engineered systems. This section derives a model that simulates unsteady-state mixed culture biofilm development under conditions of single or dual substrate limitation.

Consider the system of simultaneous biological organic carbon oxidation-nitrification. Biological nitrification itself is a two-step sequential oxidation of ammonia-nitrogen (NH_4^+-N) to nitrite-nitrogen (NO_2^--N), mediated by autotrophic bacteria such as *Nitrosomonas* spp., which is further oxidized to nitrate-nitrogen (No_3^--N) by autotrophs such as *Nitrobacter* spp. Net maximum bacterial production rates $[(\mu_m/Y) - k_e]$ for nitrifiers are significantly lower than for chemoorganotrophs oxidizing organic carbon. Consequently, in suspended culture, heterotrophic bacteria prevail in the early portions of a reactor vessel, oxidizing organics to low concentrations, and reducing heterotrophic production rates such that nitrifying bacteria can compete in the latter portions of the reactor (this presumes sufficient residence time and that essential nutrients are available). However, in a biofilm situation, microorganisms are bound to a surface, thus obviating the above residence time or growth-rate selection pressure on the two populations. A model to predict mixed bacterial culture dynamics under conditions of substrate competition must, consequently, employ a structured viewpoint.

Biofilm is considered to accumulate on the surfaces of an ideal completely mixed reactor, receiving an influent stream composed of five possible substrates, intermediates, or products (NH_4^+-N, NO_2^--N, NO_3^--N, O_2, and acetate as a model organic substrate) at an operating

* See Table 5 for Equations 17 to 25.

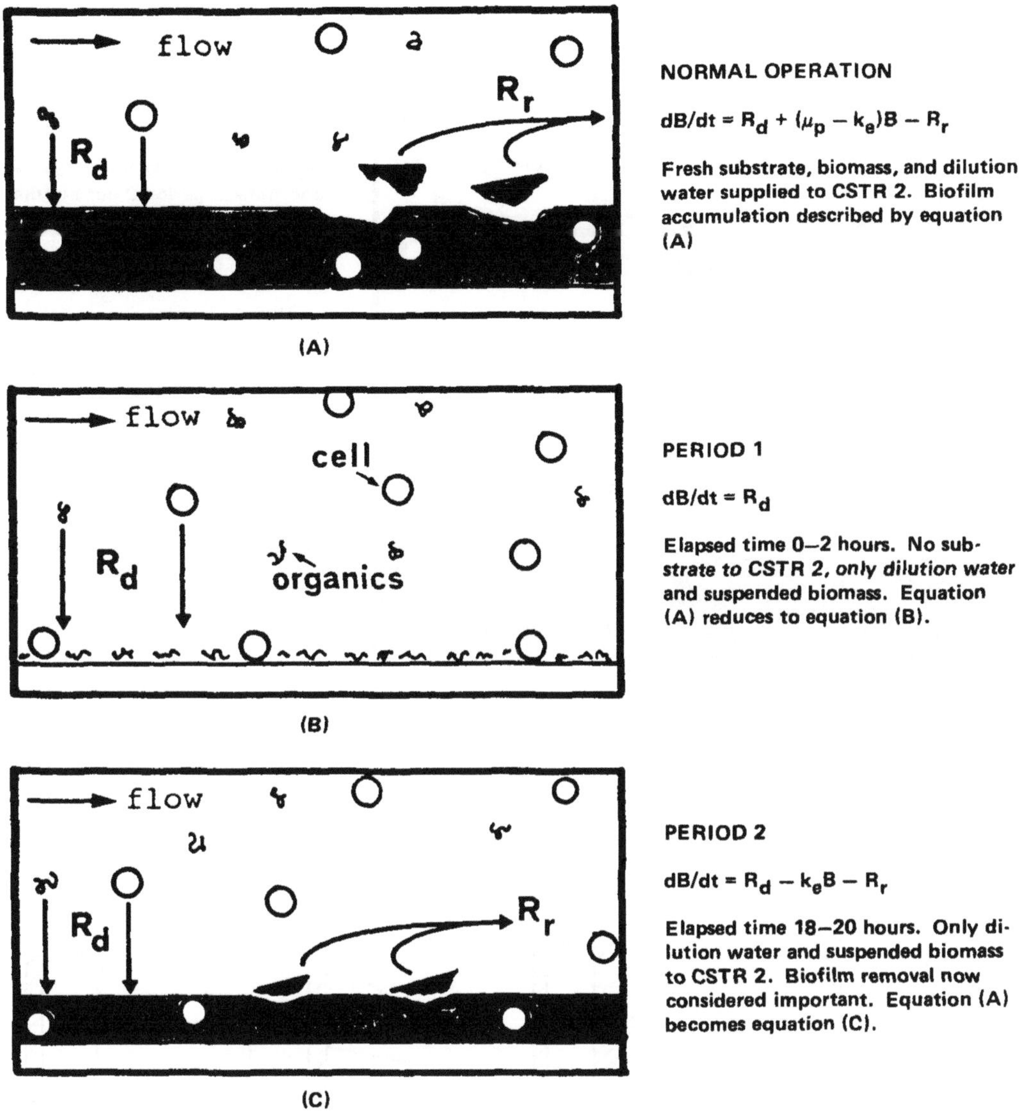

FIGURE 14. Descripton of system perturbations and simplified biofilm equations used to quantify the magnitude of individual biofilm processes.[32]

residence time that renders suspended microbial growth negligible. Assume a biofilm develops with gradients only perpendicular to the supporting media composed of the three different microbial species, heterotrophs (HET), *Nitrosomonas* spp. (NSM), and *Nitrobacter* spp. (NBC). As the biofilm develops, the diffusion path increases, thus increasing mass transfer effects within the biofilm, which in turn affects the local production rates of the different bacterial groups.

A material balance for each of the reacting substrates can be written as

$$\partial S_i(x,t)/\partial t = D_{eff}\, \partial^2 S_i(x,t)/\partial x^2 \pm SR_i(x,t) \qquad (26)$$

where $S_i(x,t)$ = local, instantaneous concentration of the "i"th substrate within the biofilm, (M_s/L^3); D_{eff} = effective diffusivity of substrate i through the biofilm, (L^2/t); and $SR_i(x,t)$ = local, instantaneous turnover rate of substrate i, (M_s/L^3t). Solution of Equation 26 for each substrate is facilitated by using the equivalent finite difference form,

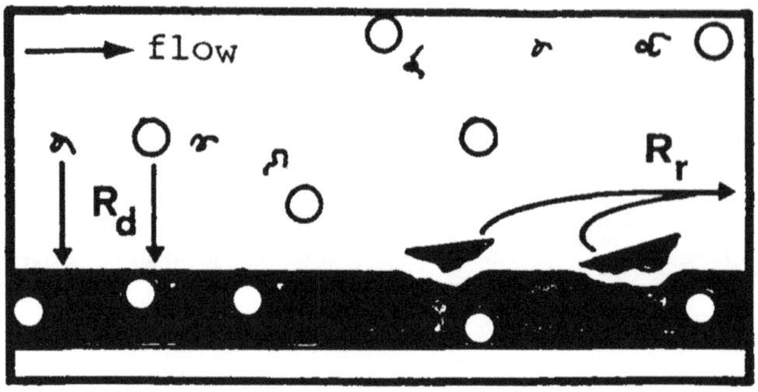

(D)

PERIOD 3

$dB/dt = R_d - k_eB - R_r$

Elapsed time 40—42 hours. Same conditions as period 2 but biofilm amount is greater.

(E)

PERIOD 4

$dB/dt = -k_eB - R_r$

Elapsed time 50—52 hours. Neither substrate nor suspended biomass to CSTR 2, only dilution water. Equation (A) reduces to equation (E).

FIGURE 14. Continued.

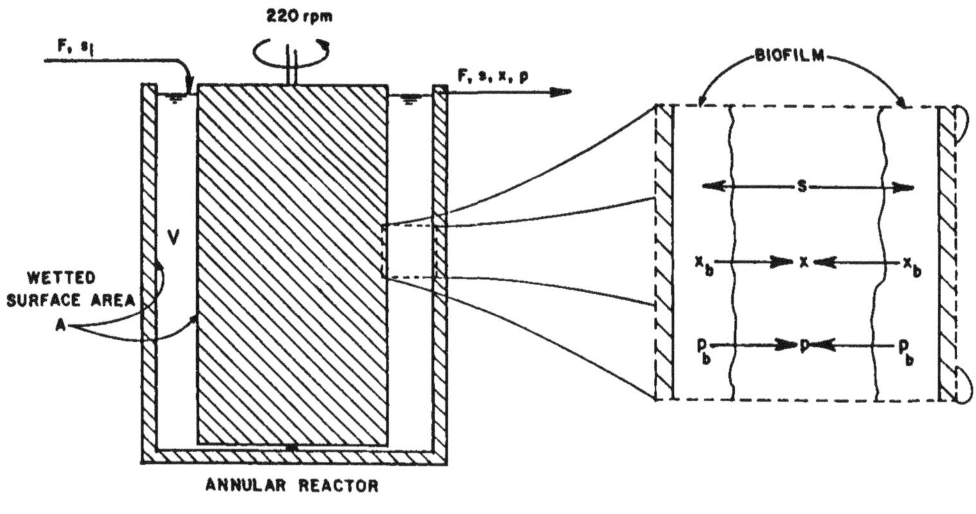

FIGURE 15. Diagram of rotating annual reactor describing structured model approach to biofilm formation.[46,47]

$$\frac{dS_{i,n}}{dt} = \frac{D_{eff}(S_{i,n+1} - 2S_{i,n} - S_{i,n-1}}{\Delta x^2} \pm SR_i \qquad (27)$$

where the substrate equations are solved over a number of spatial elements $n = 1 \rightarrow N$ at a fixed time, a time increment made, and the spatial integration repeated.

Substrate turnover rates used in Equation 26 (except that for oxygen) are of the form

Table 5
MATERIAL BALANCES AND CONSTITUTIVE EQUATIONS FOR A BIOFILM CONTINUOUS FLOW STIRRED TANK REACTOR (ANNULAR REACTOR) IN WHICH SUBSTRATE TRANSFORMATION BY SUSPENDED CELLS IS NEGLIGIBLE

Equation	Compound	Net rate of accumulation	Net rate of transport out of reactor	Net rate of transformation	Units (mass, length, time)
17	Bulk liquid substrate conc, s	$\frac{ds}{dt} =$	$D(s_i - s)$	$-(\frac{\mu}{Y_{x/s}} + \frac{q_p}{Y_{p/s}})x_b\frac{A}{B}$	ML^{-3}/t
18	Biofilm cellular areal density, x_b	$\frac{dx_b}{dt} =$		$\mu x_b - q_{dx}x_b$	ML^{-2}/t
19	Biofilm EP areal density, p_b	$\frac{dp_b}{dt} =$		$q_p x_p - q_{dp}p_b$	ML^{-2}/t
20	Suspended cellular mass, x	$\frac{dx}{dt} =$	$-Dx$	$+ q_{dx}x_b\frac{A}{V}$	ML^{-3}/t
21	Suspended EP mass, p	$\frac{dp}{dt} =$	$-Dp$	$+ q_{dp}p_b\frac{A}{V}$	ML^{-3}/t

Reaction Rate Expressions for Transformation Processes

Equation	Rate	Expression
22	Specific growth rate	$\mu = \mu_{max}\, s/(K_s + s)$
23	Specific EPS formation rate	$q_p = k\mu + k'$
24	Specific cellular detachment rate	$q_{dx} = k_{dx}x_b^n$
25	Specific EPS detachment rate	$q_{dp} = k_{dp}p_b^n$

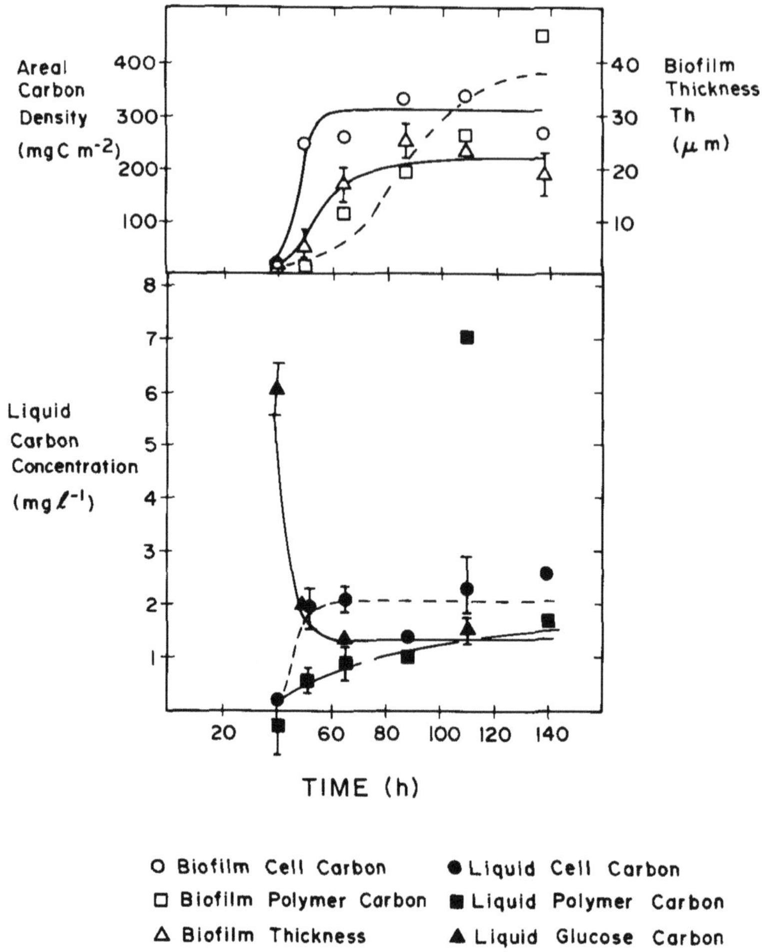

FIGURE 16. Cellular and extracellular carbon in a biofilm developed within a continuous culture of *Pseudomonas aeruginosa*.[46]

$$SR_i(x,t) = R_{i,j}(x,t)\ C_j(x,t) \tag{28}$$

where $R_{i,j}$ (x,t) = local, instantaneous removal rate of substrate i per mass of bacteria j, (M_s/M_x − t) and C_j (x,t) = local, instantaneous biofilm concentration of bacteria j, (M_x/L^3 − biofilm). Local removal rates by a specific bacterial group are assumed dual-substrate limited rate functions of both the specific electron donor and the electron acceptor (oxygen in all cases), e.g.,

$$R_{i,j}(x,t) = \frac{\hat{\mu}_{i,j}\ S_i(x,t)\ O_2(x,t)}{[Y_{j/i}]\ [K_S + S_i(x,t)]_j\ [K_{O_2} + O_2(x,t)]_j} \tag{29}$$

where O_2 = local, instantaneous oxygen concentration; $K_{S,j}$ = substrate i saturation constant for bacteria j, (M_s/L^3); $K_{O_2,j}$ = oxygen saturation constant for bacteria j, (M_{O_2}/L^3); $\mu_{i,j}$ = maximum growth rate constant for bacteria j on substrate i, (t^{-1}); $Y_{j/i}$ = stoichiometric yield for conversion of substrate i into bacterial mass j, (M_x/M_s). The oxygen turnover rate assumes that oxygen is depleted by the oxidation of the three separate exogenous substrates (NH_4^+-N, NO_2^--N, and acetate), and by the endogenous or maintenance respiration of each bacterial group.

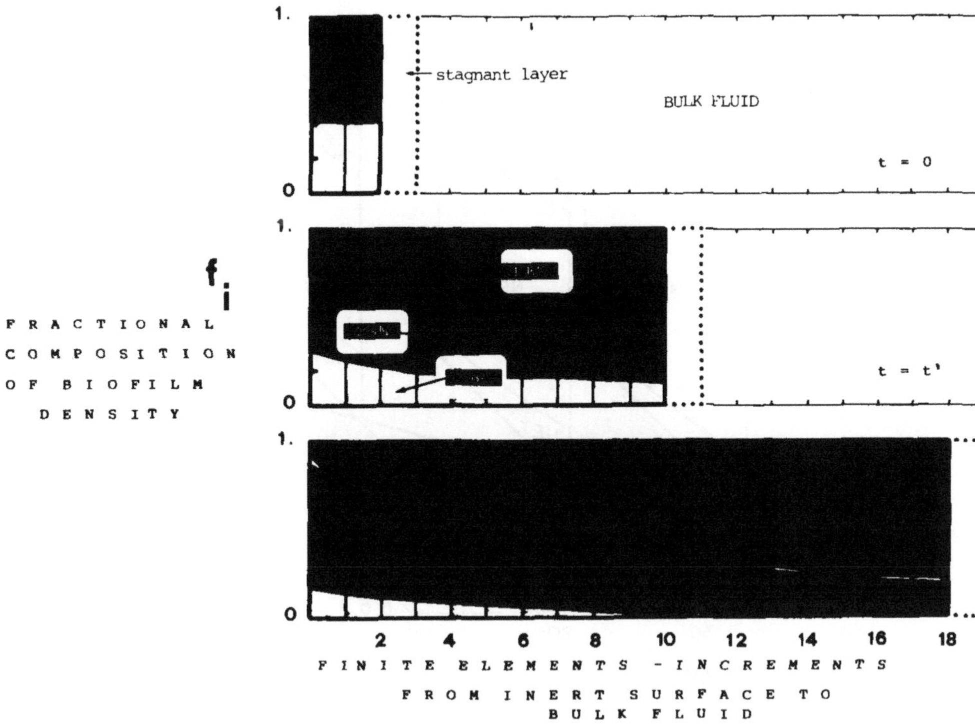

FIGURE 17. Growth of a computer-predicted mixed culture biofilm consisting of hetero- and autotrophic bacteria. Biofilm expands or contracts based upon the integral of local individual bacterial species' growth rates. See text for details of model.[97]

The change in concentration of each bacterial group within a certain element n of biofilm, over a certain time increment can be written as

$$dC_j(x,t)\ dt = BR_j(x,t) \tag{30}$$

where $BR_j(x,t)$ = local, instantaneous turnover rate for bacteria j, $(M_x/L^3/t)$. Bacterial turnover rate, within a specific element of biofilm, is the net sum of the maximum bacterial growth rate minus the endogenous decay rate. All bacterial turnover rates are of the form

$$BR_j(x,t) = (Y_{j/i})\ R_{i,j}(x,t)\ C_j(x,t) - KE_j(x,t)\ C_j(x,t) \tag{31}$$

Terms in Equation 31 remain consistent with previous definitions; the rate of endogenous decay per mass of bacteria j is defined as

$$KE_j(x,t) = \frac{K_{e,j}\ O_2(x,t)}{(K_{O_2} + O_2(x,t))} \tag{32}$$

where $K_{e,j}$ is the maximum endogenous decay rate constant for bacteria j, (t^{-1}).

Biofilm bacterial groups change spatially as a result of the turnover rate $BR_j(x,t)$ within any one biofilm volume element, n, over a time increment, dt. If within dt, a total biomass of all bacterial groups exceeds the preset biofilm density, excess mass of each bacterial group "spills-over" into the next spatial element, n + 1 (see Figure 17). As the computer simulation proceeds, the summation of individual bacterial masses in an element n must

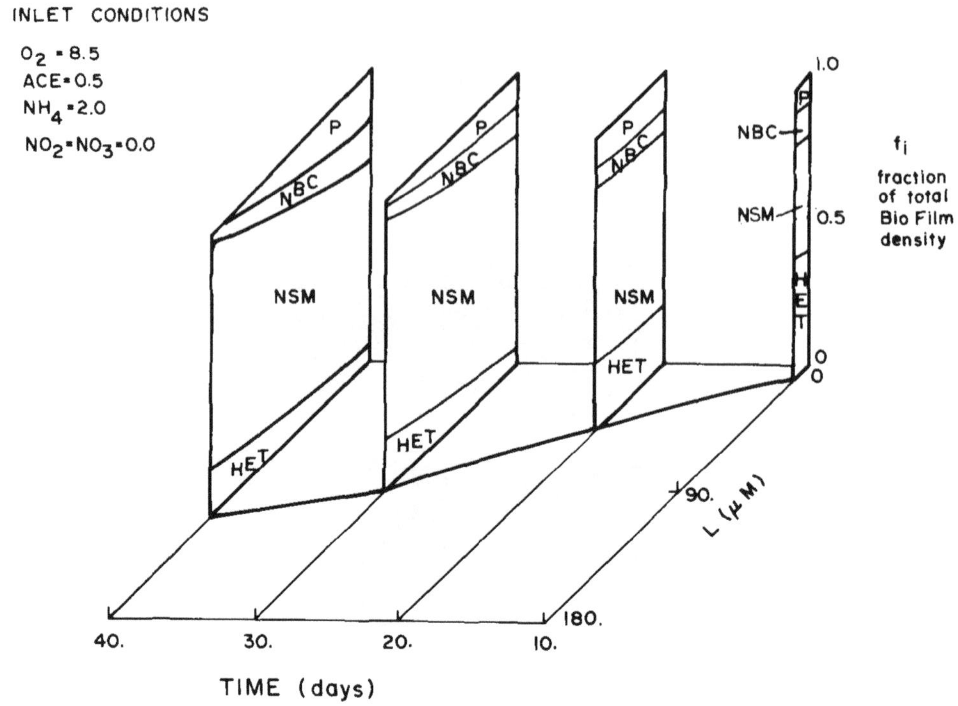

FIGURE 18. Predictions of a biofilm composed of heterotrophs (HET), *Nitrosomonas* spp. (NSM), *Nitrobacter* spp. (NBC), and extracellular polymer (P) grown under conditions that favor nitrifier dominance.[97]

include spillage from element (n − 1) to n and n to (n + 1). Accumulation of total biofilm mass (and thus biofilm thickness, since area and density are constant) proceeds at a net rate which is an integral value of all local, bacterial turnover rates. Maximum biofilm thickness is preset in the computer simulation because biofilm removal is not considered in the model. Once biofilm thickness exceeds the maximum preset limit, further growth of individual bacterial groups is "released" into the reactor fluid and appears as effluent biomass.

As the biofilm "grows", individual substrate fluxes are calculated using Equation 33

$$N_i|_{L(t)} = D_{eff}(dS_i/dx)|_{L(t)} \tag{33}$$

where L(t) = total biofilm thickness at time t, (L); and N_i = flux of substrate i in the positive x direction (note that the flux is either positive or negative depending whether the substrate is either entering or leaving the biofilm). Effluent concentrations of the five chemicals change with time as the biofilm develops. At each time increment Δt, effluent concentrations for each chemical are calculated from a material balance over the entire reactor,

$$d\bar{S}(t)/dt = D(S_i^\circ - S_i(t)) + N_i|_{L(t)} \tag{34}$$

where $\bar{S}_i$ = bulk fluid average concentration of substrate i.

Simultaneous solutions to Equations 26 through 34 are rather complicated due to the "stiffness" inherent to the equations. Kissel et al.[83] present an eloquent discussion of this problem and provides a means for numerically simulating the system equations. Examples of mixed culture biofilm development for two different sets of reactor feed conditions are given in Figures 18 and 19. For a situation where very little organic carbon is present nitrifying bacteria predominate a thin biofilm (Figure 18) while, given a moderate amount

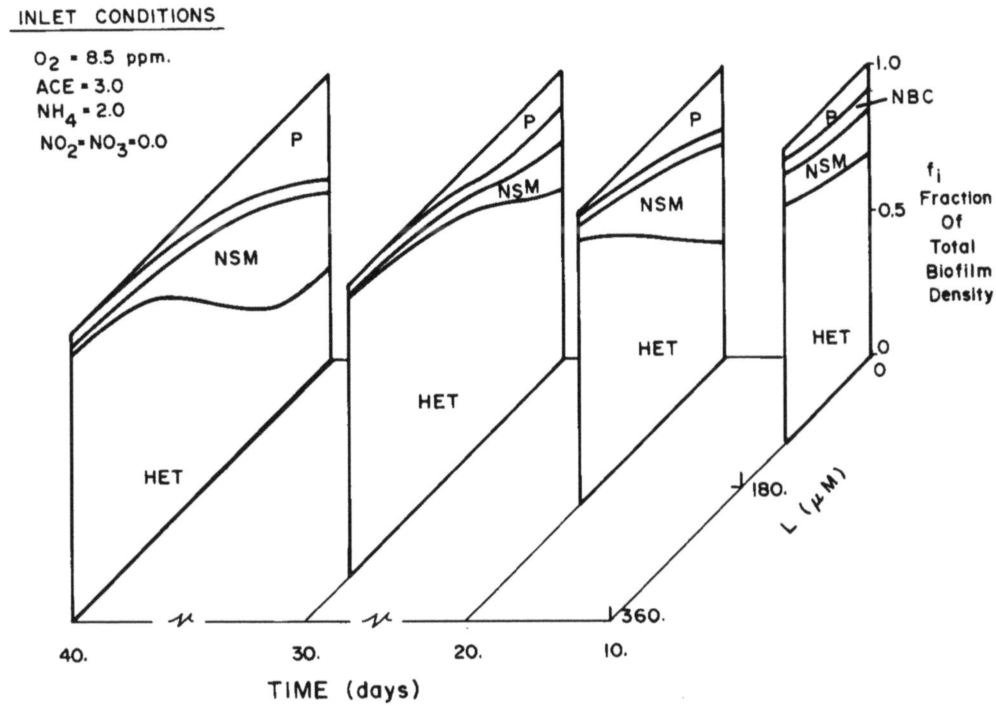

FIGURE 19. Predictions of a biofilm composed of a mixed bacterial population, as in Figure 18, grown under conditions that favor heterotroph dominance.[97]

of organic substrate, Figure 19 indicates a thick biofilm dominated by heterotrophs. Wanner and Gujer[97] use the model derived above to predict species competition and stratification in nitrifying rotating biological contactors. Kissel et al.[83] also present several simulated scenarios for biofilm development using a similar model. Tanaka and co-workers[98,99] employed a somewhat different approach, simplifying the above equations by assuming a desired bacterial composition that remained constant, then solving only the substrate equations. Harremoës[100] presents an even simpler mixed culture model of nitrification/carbon oxidation which assumes the overall growth of the biofilm to be set by the heterotrophic population only and does not consider an upper bound to overall biofilm growth.

IV. CONCLUDING REMARKS

Biofilm formation is the result of dynamic, complex phenomena where all phases of a system are intimately affected by one another through the various processes discussed in Section II. Biofilm development models and experimentation must not ignore the fact that three distinct phases will always exist and interact: (1) bulk fluid, (2) the biofilm, and (3) the substratum.

Two situations arise when considering environmental influences on biofilm formation. First, since three phases exist, then gradients will develop, which means the conditions measured in one phase may be entirely different in another part of that phase or an adjacent phase. Second, the events and conditions relating to biofilm development are time-dependent, transient even in well-controlled experimental systems.

As our needs to investigate biofilms and their development within natural and engineered systems increase, so must our techniques and analytical methods expand to quantify such spatially and temporally complex events.

REFERENCES

1. **Marshall, K. C., Ed.,** *Microbial Adhesion and Aggregation,* Springer-Verlag, Berlin, 1984.
2. **Colberg, P. C.,** personal communication.
3. **Geesey, G. G., Richardson, W. T., Yeomans, H. G., Irvin, R. T., and Costerton, J. W.,** Microscopic examination of natural sessile bacterial populations from an alpine stream, *Can. J. Microbiol.,* 23, 1733, 1977.
4. **Costerton, J. W., Geesey, G. A., and Cheng, J.-J.,** How bacteria stick, *Sci. Am.,* 238, 86, 1978.
5. **Characklis, W. G.,** Fouling biofilm development: a process analysis, *Biotechnol. Bioeng.,* 23, 1923, 1981.
6. **Characklis, W. G.,** Attached microbial growths. I. Attachment and growth, *Water Res.,* 7, 1113, 1973.
7. **Characklis, W. G.,** Attached microbial growth. II. Frictional resistances due to microbial slimes, *Water Res.,* 7, 1249, 1973.
8. **Duddridge, J. E. and Pritchard, A. M.,** Factors affecting the adhesion of bacteria to surfaces, in *Microbial Corrosion,* The Metals Society, London, 1983, 28.
9. **Marshall, K. C.,** *Interfaces in Microbial Ecology,* Harvard University Press, Cambridge, Mass., 1976.
10. **Fletcher, M. and Marshall, K. C.,** Are solid surfaces of ecological significance to aquatic bacteria, in *Advances in Microbial Ecology,* Vol. 6, Marshall, K. C., Ed., Plenum Press, New York, 1977, chap. 6.
11. **Berkeley, R. C. W., Lynch, J. M., Melling, J., Rutter, P. R., and Vincent, B., Eds.,** *Microbial Adhesion to Surfaces,* Ellis Harwood, Chichester, England, 1980.
12. **Loeb, G. and Neihoff, R.,** Marine conditioning films, in *Applied Chemistry at Protein Interfaces,* Advances in Chemistry Ser. No. 145, Baier, R. E., Ed., American Chemical Society, Washington, D.C., 1973, 319.
13. **Baier, R. E., Shafin, E. G., and Zisman, W. A.,** Adhesion: mechanisms that assist or impede it, *Science,* 162, 1360, 1968.
14. **Baier, R. E. and Weiss, L.,** Demonstration of the involvement of adsorbed proteins in cell adhesion and cell growth on solid surfaces, in *Applied Chemistry at Protein Interfaces,* Advances in Chemistry, No. 145, American Chemical Society, Washington, D. C., 1975, 300.
15. **Baier, R. E. and Depalma, V. A.,** Microfouling on Metallic and Coated Metallic Flow Surfaces in Model Heat Exchanger Cells, Report by Calspan Corp., Buffalo, N. Y., 1977.
16. **Baier, R. E.,** Influences of the initial surface condition of materials on bioadhesion, in *Proc. 3rd Int. Congr. Marine Corrosion and Biofouling,* National Bureau of Standards, Gaithersburg, Md., 1972.
17. **Dexter, S. C., Sullivan, J. D., Williams, J., and Watson, J.,** Influence of substratum wettability on the attachment of marine bacteria to wetted surfaces, *Appl. Microbiol.,* 30, 298, 1975.
18. **Dexter, S. C.,** Influence of substratum wettability on the formation of bacterial slime films on solid surfaces immersed in natural seawater, in *Proc. 4th Int. Congr. Marine Corrosion and Biofouling,* National Bureau of Standards, Gaithersburg, Md., 1976.
19. **Dexter, S. C.,** Some new possibilities for biofouling control, in *Proc. OTEC Biofouling and Corrosion Symp.,* Seattle, Washington, 1977.
20. **Fletcher, M. and Loeb, G. I.,** Influence of substratum characteristics on the attachment of a marine pseudomonad to solid surfaces, *Appl. Environ. Microbiol.,* 37, 67, 1979.
21. **Valentine, R. C. and Allison, A. C.,** Virus particles adsorption: theory of adsorption and experiments on attachment of particles to nonbiological surfaces, *Biochem. Biophys. Acta,* 34, 10, 1959.
22. **Corpe, W. A.,** Attachment of marine bacteria to solid surfaces, in *Adhesion in Biological Systems,* Manly, R. S., Ed., Academic Press, New York, 1970.
23. **Tosteson, T. R. and Corpe, W. A.,** Enhancement of adhesion of the marine *Chlorella vulgaris* to glass, *Am. J. Microbiol.,* 21, 1025, 1975.
24. **Zaidi, B. R., Bard, R. F., and Tosteson, T. R.,** Microbial specificity of metallic surfaces exposed to ambient seawater, *Appl. Environ. Microbiol.,* 48, 519, 1984.
25. **Imam, S. H., Bard, R. F., and Tosteson, T. R.,** Specificity of marine microbial surface interactions, *Appl. Environ. Microbiol.,* 48, 833, 1984.
26. **LaMotta, E. J., Hickey, R. F., and Buydos,** Effect of polyelectrolytes on biofilm growth, *J. Environ. Eng. Div., ASCE,* 108(EE6), 1982.
27. **Bryers, J. D.,** Dynamics of Early Biofilm Formation in a Turbulent Flow System, Ph.D. dissertation, Rice University, Houston, 1980.
28. **Fletcher, M.,** The effects of culture concentration and age, time, and temperature on bacterial attachment to polystyrene, *Can. J. Microbiol.,* 23, 1, 1977.
29. **Bowen, B. D., Levine, S., and Epstein, N. J.,** *J. Colloid Interface Sci.,* 54, 375, 1976.
30. **Powell, M. S. and Slater, N. K. H.,** The deposition of bacterial cells to solid surfaces, *Biotechnol. Bioeng.,* 25, 891, 1983.
31. **Beal, S. K.,** Deposition of particles in turbulent flow on channel or pipe walls, *Nucl. Sci. Eng.,* 40, 1, 1970.
32. **Bryers, J. D. and Characklis, W. G.,** Processes governing primary biofilm formation, *Biotechnol. Bioeng.,* 24, 2451, 1982.

33. **Duddridge, J. E., Kent, C. A., and Laws, J. F.,** Effect of surface shear stress on the attachment of *Pseudomonas fluorescens* to stainless steel under defined flow conditions, *Biotechnol. Bioeng.*, 24, 153, 1982.
34. **Marshall, K. C., Stout, R., and Mitchell, R.,** Mechanism of the initial events in the sorption of marine bacteria at interfaces, *J. Gen. Microbiol.*, 68, 337, 1971.
35. **Fletcher, M. and Floodgate, G. D.,** An electron-microscopic demonstration of an acidic polysaccharide involved in adhesion of a marine bacterium to solid surfaces, *J. Gen. Microbiol.*, 74, 325, 1973.
36. **Marshall, K. C. and Cruickshank, N. H.,** Cell surface hydrophobicity and the orientation of certain bacteria at surfaces, *Arch. Mikrobiol.*, 91, 29, 1973.
37. **Fletcher, M.,** The question of passive vs. active attachment mechanisms in nonspecific bacterial adhesion, in *Microbial Adhesion to Surfaces*, Berkeley, R. C. W., Lynch, J. M., Melling, J., Rutter, P. R., and Vincent, B., Eds., Ellis Harwood, Chichester, England, 1980, 197.
38. **Sutherland, I. W.,** Polysaccharides in the adhesion of marine and freshwater bacteria, in *Microbial Adhesion to Surfaces*, Berkeley, R. C. W., Lynch, J. M., Melling, J., Rutter, P. R., and Vincent, B., Eds., Ellis Harwood, Chichester, England, 1980, 197.
39. **Corpe, W. A., Matsuuchi, L., and Armbruster, B.,** Secretion of adhesive polymers and attachment of marine bacteria to surfaces, in *Proc. 3rd Int. Biodegradation Symp.*, Sharpley, J. M., and Kaplan, A. M., Eds., Applied Science, London, 1976, 433.
40. **Rutter, P. R. and Vincent, B.,** The adhesion of microorganisms to surfaces: physio-chemical aspects, in *Microbial Adhesion to Surfaces*, Berkeley, R. C. W., Lynch, J. M., Melling, J., Rutter, P. R., and Vincent, B., Eds., Ellis Harwood, Chichester, England, 1980, 79.
41. **Marshall, K. C., Stout, R., and Mitchel, R.,** Selective sorption of bacteria from seawater, *Can. J. Microbiol.*, 17, 1413, 1971.
42. **Characklis, W. G.,** Biofilm development: a process analysis, in *Microbial Adhesion and Aggregation*, Marshall, K. C., Ed., Springer-Verlag, Berlin, 1984, 137.
43. **Baltzis, B. C. and Fredrickson, A. G.,** Competition of two microbial populations for a simple resource in a chemostat when one of them exhibits wall attachment, *Biotechnol. Bioeng.*, 25, 2419, 1983.
44. **Dabros, T. and van de Ven, T. G. M..** A direct method for studying particle deposition onto solid surfaces, *Colloid Polym. Sci.*, 261, 694, 1983.
45. **Filip, Z. and Hattori, T.,** Utilization of substrates and transformation of solid substrata, in *Microbial Adhesion and Aggregation*, Marshall, K. C., Ed., Springer-Verlag, Berlin, 1984, 251.
46. **Trulear, M. G.,** Cellular Reproduction and Extracellular Polymer Formation in the Development of Biofilms, Ph.D. dissertation, Montana State University, Bozeman, 1983.
47. **Bakke, R., Trulear, M. G., Robinson, J. A., and Characklis, W. G.,** Activity of *Pseudomonas aeruginosa* in biofilms: steady-state, *Biotechnol. Bioeng.*, 26, 1418, 1984.
48. **Andrews, G. F. and Tien, C.,** Bacterial film growth in adsorbent surfaces, *AIChE. J.*, 27, 396, 1981.
49. **Atkinson, B. and Howell, J. A.,** Slime holdup, influent BOD, and mass transfer in trickling filters, *J. Environ. Eng. Div. ASCE*, 101, 585, 1975.
50. **Atkinson, B., Busch, A. W., Swilley, E. L., and Williams, O. A.,** Kinetic, mass transfer, and organism growth in a biological film reactor, *Trans. Inst. Chem. Eng.*, 45, 257, 1976.
51. **Williamson, K. and McCarty, P. L.,** A model of substrate utilization by bacterial films, *J. Water Pollut. Control Fed.*, 48, 9, 1976.
52. **Kornegay, B. H. and Andres, J. F.,** Kinetics of fixed film biological reactors, *J. Water Pollut. Control Fed.*, 40, R460, 1968.
53. **Atkinson, B. and Daoud, I. S.,** Diffusion effects with(in) microbial films, *Trans. Inst. Chem. Eng.*, 48, 245, 1970.
54. **Atkinson, B. and How, S. Y.,** The overall rate of substrate uptake by microbial films. II. Effects of concentration and thickness with mixed microbial films, *Trans. Inst. Chem. Eng.*, 52, 260, 1974.
55. **Harremoes, P.,** Half order reactions in biofilm and filter kinetics, *Vatten*, 33, 122, 1977.
56. **Harremoes, P.,** Biofilm kinetics, in *Water Pollution Microbiology*, Vol. 2, Mitchell, R., Ed., John Wiley & Sons, New York, 1978, 82.
57. **Riemer, M. and Harremoes, P.,** Multi-component diffusion in denitrifying biofilms, *Prog. Water Technol.*, 10, 149, 1978.
58. **Rittmann, B. E. and Brunner, C. W.,** The nonsteady-state biofilm process for advanced organics removal: concept, model, and experimental evaluation, presented at Annu. Conf. Water Pollution Control Fed., Atlanta, October 3, 1983.
59. **Rittmann, B. E. and McCarty, P. L.,** Substrate flux into biofilms of any thickness, *J. Environ. Eng. Div. ASCE*, 107 (EE4), 831, 1981.
60. **Grady, C. P. L.,** Modeling of biological fixed films — a state of the art review, *In Proc. 1st Int. Conf. Fixed-Film Biological Processes*, University of Pittsburg Press, Pittsburg, Pa., Wu, Y. C., Smith, E. D., Miller, R. D., and Patken, E. J. O., Eds., April 20, 1982, Kings Island, Ohio.

61. **Atkinson, B. and Abdel Ralmen Ali, M. E.,** The effectiveness of biomass hold-up and packing surface in trickling filters, *Water Res.*, 12, 147, 1978.
62. **Grady, C. P. L., Jr. and Lim, H. C.,** *Biological Wastewater Treatment Theory and Application*, Marcel Dekker, New York, 1980, chap. 14.
63. **Grady, C. P. L., Jr. and Lim, H. C.,** A conceptual model of RBC performance, in *Proc. 1st Natl. Symp. RBC Technol.*, Vol. 2, University of Pittsburg Press, Pittsburg, Pa., Smith, E. D., Miller, R. D., and Wu, Y. C., Eds., 1980, 829.
64. **Howell, J. A. and Atkinson, B.,** Sloughing of microbial film in trickling filters, *Water Res.*, 10, 307, 1976.
65. **Rittmann, B. E. and McCarty, P. L.,** Model of steady-state biofilm kinetics, *Biotechnol. Bioeng.*, 22, 2343, 1980.
66. **Rittmann, B. E. and McCarty, P. L.,** Variable order model of bacterial film kinetics, *J. Environ. Eng. Div. ASCE*, 104, 889, 1978.
67. **Jennings, P. A., Snoeyink, V. L., and Chian, E. S. K.,** Theoretical model for a submerged biological filter, *Biotechnol. Bioeng.*, 18, 1249, 1976.
68. **Wang, S.-C. P.,** The Interaction Between Adsorption and Microbial Growth in Biological Activated Carbon Process, Ph.D. thesis, Syracuse University, Syracuse, N.Y., 1981.
69. **Mulcahy, L. T., Shieh, W. K., and LaMotta, E. J.,** Simplified mathematical models for fluidized bed biofilm reactor, in *Water-1980*, AICh.E. Symp. Ser., No. 77, American Institute of Chemical Engineers, Washington, D.C., 1981, 273.
70. **Shieh, W. K.,** A suggested kinetic model for the fluidized bed biofilm reactor, *Biotechnol. Bioeng.*, 22, 667, 1980.
71. **Mulcahy, L. T., Shieh, W. K., and LaMotta, E. J.,** Kinetic model of biological denitrification in a fluidized bed biofilm reactor, *Prog. Water Technol.*, 12, 143, 1980.
72. **Meunier, A. D. and Williamson, K. J.,** Packed bed biofilm reactors, simplified model, *J. Environ. Eng. Div. ASCE*, 107, 307, 1981.
73. **Williamson, K. J. and Chung, T. H.,** Dual limitation of substrate utilization kinetics within bacterial films, presented at 49th Meet. Am. Inst. Chem. Eng., Houston, March 19, 1975.
74. **Janson, J. la C. and Kristensen, G. H.,** Fixed Film Kinetics — Denitrification in Fixed Films, Report No. 80-59, Department of Environmental Engineering, Technical University of Denmark, Copenhagen, 1980.
75. **Howell, J. A. and Atkinson, B.,** Influence of oxygen and substrate concentrations on the ideal film thickness and maximum overall substrate uptake rate in microbial film fermenters, *Biotechnol. Bioeng.*, 18, 15, 1976.
76. **Harris, N. P. and Hansford, G. S.,** A study of substrate removal in a microbial film reactor, *Water Res.*, 10, 935, 1976.
77. **Famularo, J., Mueller, J. A., and Mulligan, T.,** Application of mass transfer to rotating biological contactors, *J. Water Pollut. Control Fed.*, 50, 653, 1978.
78. **Mueller, J. A., Paquin, P., and Famularo, J.,** Nitrification in rotating biological contactors, *J. Water Pollut. Control Fed.*, 52, 688, 1980.
79. **Mueller, J. A., Paquin, P., and Famularo, J.,** Mass transfer impact on RBC and trickling filter design, *Water-1979*, AICh.E Symp. Ser. No. 76, American Institute of Chemical Engineers, Washington, D.C., 1980, 250.
80. **Aiegrist, H.,** Stofftransportprozesse in Festsitzender Biomass, Ph.D. dissertation, Eidg. Technische Hochschüle, Zurich, 1985.
81. Mechanisms of Adhesion, Group One Report, in *Microbial Adhesion and Aggregation*, Marshall, K. C., Ed., Springer-Verlag, Berlin, 1984, 5.
82. **Hirokawa, Y., Tanaka, T., and Katayama, S.,** Effects of network structure on the phase transition of acrylamide-sodium acrylate copolymer gels, in *Microbial Adhesion and Aggregation*, Marshall, K. C., Ed., Springer-Verlag, Berlin, 1984, 177.
83. **Kissel, J. C., McCarty, P. L., and Street, R. L.,** Numerical simulation of (a) mixed culture biofilm, *J. Environ. Eng. Div. ASCE*, 110, 393, 1983.
84. **Trulear, M. G.,** Rate of Biofilm Development in a Turbulent Flow Field, M.S. thesis, Rice University, Houston, 1979.
85. **Rittmann, B. E.,** The effect of shear stress on biofilm loss rate, *Biotechnol. Bioeng.*, 24, 501, 1983.
86. **Zelver, N.,** Biofilm Development and Associated Energy Losses in Water Conduits, M.S. thesis, Rice University, Houston, 1979.
87. **Bryers, J.,** Application of captured cell systems in biological treatment, in *Bioenvironmental Systems*, Vol. 1, Wise, D. L., Ed., CRC Press, Boca Raton, Fla., 1987, 27.
88. **Bryers, J. D. and Egli, Th.,** Anaerobic bacterial degradation of NTA in a chemostat study, unpublished data.

89. **Hawkes, H. A. and Shephard, M. R. N.,** The seasonal accumulation of solids in percolating filters and attempted control at low frequency dosing, in Proc. 5th Int. Water Pollution Res. Conf., paper II-11/1-8, 1970.
90. **Heulekian, H. and Crosby, E. S.,** Slime formation in polluted waters, *Sewage Ind. Wastes,* 28, 78, 1956.
91. **Harremoës, P., Jansen, J. la Cour, and Kristensen, G. H.,** Practical problems related to nitrogen bubble formation in fixed film reactors, *Prog. Water Technol.,* 12, 253, 1980.
92. **Arvin, E. and Kristensen, G.H.,** Effect of denitrification on the pH in biofilms, *Water Sci. Technol.,* 14, 833, 1982.
93. **Bakke, R. and Characklis, W. G.,** personal communication. 1984.
94. **Roels, R. A. and Kossen, N. W. F.,** On the modeling of microbial metabolism, in *Progress in Industrial Microbiology,* Vol. 14, Bull, J. J., Ed., Elsevier, Amsterdam, 1978, 95.
95. **Bryers, J. D.,** Biofilm formation and chemostat dynamics: pure and mixed culture considerations, *Biotechnol. Bioeng.,* 26, 948, 1984.
96. **Luedeking, R. and Piret, E. L.,** A kinetic study of the lactic acid fermentation, *J. Biochem. Microbiol. Technol. Eng.,* 1, 893, 1959.
97. **Wanner, O. and Gujer, W.,** Competition in biofilms, *Water Sci. Technol.,* 17, 27, 1983.
98. **Tanaka, H., Uzman, S., and Dunn, I. J.,** Kinetics of nitrification using a fluidized sand bed reactor with attached growth, *Biotechnol. Bioeng.,* 23, 1683, 1981.
99. **Tanaka, H. and Dunn, I. J.,** Kinetics of biofilm nitrification, *Biotechnol. Bioeng.,* 24, 669, 1982.
100. **Harremoës, P.,** Criteria for nitrification in fixed film reactors, *Water Sci. Technol.,* 14, 167, 1982.
101. **Escher, A. R.,** Colonization of a smooth surface by *Pseudomonas aeruginosa:* Image Analysis Methods, Ph.D. thesis, Montana State University Press, Bozeman, 1986.

INDEX

A

B

C

D

N

O

P

R

S

T

U

V

W

Z

For Product Safety Concerns and Information please contact our EU representative GPSR@taylorandfrancis.com
Taylor & Francis Verlag GmbH, Kaufingerstraße 24, 80331 München, Germany

www.ingramcontent.com/pod-product-compliance
Ingram Content Group UK Ltd.
Pitfield, Milton Keynes, MK11 3LW, UK
UKHW052053230926
475702UK00025B/1150
9781315896540